EMMY NOETHER

PURE AND APPLIED MATHEMATICS

A Program of Monographs, Textbooks, and Lecture Notes

EXECUTIVE EDITORS—MONOGRAPHS, TEXTBOOKS, AND LECTURE NOTES

Earl J. Taft
Rutgers University
New Brunswick, New Jersey

Edwin Hewitt
University of Washington
Seattle, Washington

CHAIRMAN OF THE EDITORIAL BOARD

S. Kobayashi
University of California, Berkeley
Berkeley, California

EDITORIAL BOARD

Glen E. Bredon
Rutgers University

Irving Reiner
University of Illinois at Urbana-Champaign

Sigurdur Helgason
Massachusetts Institute of Technology

Fred S. Roberts
Rutgers University

Marvin Marcus
University of California, Santa Barbara

Paul J. Sally, Jr.
University of Chicago

W. S. Massey
Yale University

Jane Cronin Scanlon
Rutgers University

Zuhair Nashed
University of Delaware

Martin Schechter
Yeshiva University

Donald Passman
University of Wisconsin

Julius L. Shaneson
Rutgers University

Olga Taussky Todd
California Institute of Technology

MONOGRAPHS AND TEXTBOOKS IN
PURE AND APPLIED MATHEMATICS

1. *K. Yano*, Integral Formulas in Riemannian Geometry (1970) *(out of print)*
2. *S. Kobayashi*, Hyperbolic Manifolds and Holomorphic Mappings (1970) *(out of print)*
3. *V. S. Vladimirov*, Equations of Mathematical Physics (A. Jeffrey, editor; A. Littlewood, translator) (1970)
4. *B. N. Pshenichnyi*, Necessary Conditions for an Extremum (L. Neustadt, translation editor; K. Makowski, translator) (1971)
5. *L. Narici, E. Beckenstein, and G. Bachman*, Functional Analysis and Valuation Theory (1971)
6. *D. S. Passman*, Infinite Group Rings (1971)
7. *L. Dornhoff*, Group Representation Theory (in two parts). Part A: Ordinary Representation Theory. Part B: Modular Representation Theory (1971, 1972)
8. *W. Boothby and G. L. Weiss (eds.)*, Symmetric Spaces: Short Courses Presented at Washington University (1972)
9. *Y. Matsushima*, Differentiable Manifolds (E. T. Kobayashi, translator) (1972)
10. *L. E. Ward, Jr.*, Topology: An Outline for a First Course (1972) *(out of print)*
11. *A. Babakhanian*, Cohomological Methods in Group Theory (1972)
12. *R. Gilmer*, Multiplicative Ideal Theory (1972)
13. *J. Yeh*, Stochastic Processes and the Wiener Integral (1973) *(out of print)*
14. *J. Barros-Neto*, Introduction to the Theory of Distributions (1973) *(out of print)*
15. *R. Larsen*, Functional Analysis: An Introduction (1973)
16. *K. Yano and S. Ishihara*, Tangent and Cotangent Bundles: Differential Geometry (1973)
17. *C. Procesi*, Rings with Polynomial Identities (1973)
18. *R. Hermann*, Geometry, Physics, and Systems (1973)
19. *N. R. Wallach*, Harmonic Analysis on Homogeneous Spaces (1973)
20. *J. Dieudonné*, Introduction to the Theory of Formal Groups (1973)
21. *I. Vaisman*, Cohomology and Differential Forms (1973)
22. *B.-Y. Chen*, Geometry of Submanifolds (1973)
23. *M. Marcus*, Finite Dimensional Multilinear Algebra (in two parts) (1973, 1975)
24. *R. Larsen*, Banach Algebras: An Introduction (1973)
25. *R. O. Kujala and A. L. Vitter (eds)*, Value Distribution Theory: Part A; Part B. Deficit and Bezout Estimates by Wilhelm Stoll (1973)
26. *K. B. Stolarsky*, Algebraic Numbers and Diophantine Approximation (1974)
27. *A. R. Magid*, The Separable Galois Theory of Commutative Rings (1974)
28. *B. R. McDonald*, Finite Rings with Identity (1974)
29. *J. Satake*, Linear Algebra (S. Koh, T. Akiba, and S. Ihara, translators) (1975)

30. *J. S. Golan*, Localization of Noncommutative Rings (1975)
31. *G. Klambauer*, Mathematical Analysis (1975)
32. *M. K. Agoston*, Algebraic Topology: A First Course (1976)
33. *K. R. Goodearl*, Ring Theory: Nonsingular Rings and Modules (1976)
34. *L. E. Mansfield*, Linear Algebra with Geometric Applications: Selected Topics (197
35. *N. J. Pullman*, Matrix Theory and Its Applications (1976)
36. *B. R. McDonald*, Geometric Algebra Over Local Rings (1976)
37. *C. W. Groetsch*, Generalized Inverses of Linear Operators: Representation and Approximation (1977)
38. *J. E. Kuczkowski and J. L. Gersting*, Abstract Algebra: A First Look (1977)
39. *C. O. Christenson and W. L. Voxman*, Aspects of Topology (1977)
40. *M. Nagata*, Field Theory (1977)
41. *R. L. Long*, Algebraic Number Theory (1977)
42. *W. F. Pfeffer*, Integrals and Measures (1977)
43. *R. L. Wheeden and A. Zygmund*, Measure and Integral: An Introduction to Real Analysis (1977)
44. *J. H. Curtiss*, Introduction to Functions of a Complex Variable (1978)
45. *K. Hrbacek and T. Jech*, Introduction to Set Theory (1978)
46. *W. S. Massey*, Homology and Cohomology Theory (1978)
47. *M. Marcus*, Introduction to Modern Algebra (1978)
48. *E. C. Young*, Vector and Tensor Analysis (1978)
49. *S. B. Nadler, Jr.*, Hyperspaces of Sets (1978)
50. *S. K. Sehgal*, Topics in Group Rings (1978)
51. *A. C. M. van Rooij*, Non-Archimedean Functional Analysis (1978)
52. *L. Corwin and R. Szczarba*, Calculus in Vector Spaces (1979)
53. *C. Sadosky*, Interpolation of Operators and Singular Integrals: An Introduction to Harmonic Analysis (1979)
54. *J. Cronin*, Differential Equations: Introduction and Quantitative Theory (1980)
55. *C. W. Groetsch*, Elements of Applicable Functional Analysis (1980)
56. *I. Vaisman*, Foundations of Three-Dimensional Euclidean Geometry (1980)
57. *H. I. Freedman*, Deterministic Mathematical Models in Population Ecology (1980)
58. *S. B. Chae*, Lebesgue Integration (1980)
59. *C. S. Rees, S. M. Shah, and Č. V. Stanojević*, Theory and Applications of Fourier Analysis (1981)
60. *L. Nachbin*, Introduction to Functional Analysis: Banach Spaces and Differential Calculus (R. M. Aron, translator) (1981)
61. *G. Orzech and M. Orzech*, Plane Algebraic Curves: An Introduction Via Valuations (1981)
62. *R. Johnsonbaugh and W. E. Pfaffenberger*, Foundations of Mathematical Analysis (1981)

63. *W. L. Voxman and R. H. Goetschel,* Advanced Calculus: An Introduction to Modern Analysis (1981)
64. *L. J. Corwin and R. H. Szczarba,* Multivariable Calculus (1982)
65. *V. I. Istrățescu,* Introduction to Linear Operator Theory (1981)
66. *R. D. Järvinen,* Finite and Infinite Dimensional Linear Spaces: A Comparative Study in Algebraic and Analytic Settings (1981)
67. *J. K. Beem and P. E. Ehrlich,* Global Lorentzian Geometry (1981)
68. *D. L. Armacost,* The Structure of Locally Compact Abelian Groups (1981)
69. *J. W. Brewer and M. K. Smith, editors,* Emmy Noether: A Tribute to Her Life and Work (1981)

Other Volumes in Preparation

EMMY NOETHER
A Tribute to Her Life and Work

EDITORS

James W. Brewer
The University of Kansas
Lawrence, Kansas

Martha K. Smith
The University of Texas at Austin
Austin, Texas

MARCEL DEKKER, INC. New York and Basel

Library of Congress Cataloging in Publication Data
Main entry under title:

Emmy Noether: a tribute to her life and work.

  Bibliography: p.
  Includes index.
  1. Noether, Emmy, 1882-1935. 2. Mathematicians—Germany—Biography. 3. Algebra—Addresses, essays, lectures. I. Noether, Emmy, 1882-1935. II. Brewer, James W., [date]. III. Smith, Martha K., [date].
QA29.N6E47   510'.92'4 [B]   81-15203
ISBN 0-8247-1550-0   AACR2

COPYRIGHT © 1981 by MARCEL DEKKER, INC. ALL RIGHTS RESERVED.

Neither this book nor any part may be reproduced or transmitted in any form or by any means, electronic or mechanical, including photocopying, microfilming, and recording, or by any information storage and retrieval system, without permission in writing from the publisher.

MARCEL DEKKER, INC.
270 Madison Avenue, New York, New York 10016

Current printing (last digit);
10 9 8 7 6 5 4 3 2 1

PRINTED IN THE UNITED STATES OF AMERICA

Preface

Although this volume was originally conceived to commemorate Emmy Noether's one hundredth birthday, we feel that it will fulfill several other functions as well. Interest in Noether and her mathematics has been high in recent years, and for several reasons. First, there has been a general interest in the lives of the mathematicians who were major contributors to the development of modern mathematics, as witness Constance Reid's books on Hilbert and Courant, and the popularity of Springer's posters of famous mathematicians. This historical interest is no doubt linked to a renewed interest in the work of these people, as so many branches of mathematics are returning in part to the techniques and questions studied at the beginning of this century. Finally, there is an additional interest in Noether as one of the first women to achieve high distinction in mathematics. We modestly hope that this book will serve all of these interests.

The book progresses from the primarily biographical to the primarily mathematical. The first section consists of a biography of Emmy Noether. It should be accessible to readers with little mathematical background. The only previous comprehensive biography is that by Auguste Dick, which has only recently been translated into English. The second section mixes the biographical and the mathematical. Most of it does not require a great deal of mathematical background, but is probably of interest mainly to mathematicians and historians of science. We include here English translations of Alexandroff's and van der Waerden's obituaries of Noether. (The third major memorial, by H. Weyl, is not included since it originally appeared in English and is easily accessible.) The third section consists of a series of articles on certain areas in which Noether worked. Each one is written by a mathematician active in that area. The level of exposition of these articles is roughly that of the *American Mathematical Monthly*. We feel that they will be useful to teachers, graduate students, and historians of mathematics, as well as to research mathematicians. The final section consists of an English translation of Noether's address to the 1932 International Congress of Mathematicians.

A complete list of the publications of Emmy Noether is included at the end of the book. In all chapters, numbered references refer to this list. References to works by other authors are cited by name and date and are listed at the end of each chapter.

We wish to thank all the contributors to the book for their cooperation (especially Clark Kimberling, who was involved with the project before we were); Earl Taft for his long involvement in the idea of a memorial volume on Emmy Noether; the University of Kansas for clerical assistance; and the University of Texas for the use of its libraries.

J. W. Brewer
M. K. Smith

Contents

Preface   iii

Contributors   ix

I   BIOGRAPHY

     1. Emmy Noether and Her Influence

       Clark Kimberling

       Family History / Max Noether and Paul Gordan / Childhood and Education (1882-1908) / The First Epoch (1908-1919) / The Second Epoch (1920-1926) / The Third Epoch (1927-1935) / In America (1933-1935) / Students, Colleagues, and Influence / Notes / Acknowledgments / References

II   NOETHER AND HER COLLEAGUES

     2. Mathematics at the University of Göttingen (1931-1933)   65

       Saunders Mac Lane

       Appendix A: Mathematicians at Gottingen 1931-1933 / Appendix B: A View of the Lectures in Mathematics at Gottingen / References

3. My Personal Recollections of Emmy Noether 79
   Olga Taussky
   References

4. Obituary of Emmy Noether 93
   B. L. van der Waerden

5. In Memory of Emmy Noether 99
   P. S. Alexandroff
   Notes

III NOETHER'S MATHEMATICS

6. Galois Theory 115
   Richard G. Swan
   1. Introduction / 2. Constructing Galois Extensions / 3. Generic Galois Extensions / 4. Fischer's Theorem / 5. Galois Descent / 6. A Birational Invariant / 7. Stable Equivalence / 8. Lenstra's Theorem / 9. Tori / 10. Steenrod's Problem / References

7. The Calculus of Variations 125
   E. J. McShane

8. Commutative Ring Theory 131
   Robert Gilmer
   Idealtheorie in Ringbereichen / *Abstrakter Aufbau der Idealtheorie in algebraischen Zahl- und Funktionenkörpern* / Emmy Noether's Influence in Commutative Ring Theory / Acknowledgments

9. Representation Theory 145
   T. Y. Lam
   Notes / References

10. Algebraic Number Theory 157
    A. Fröhlich
    Galois Module Structure / Cohomology, or Central Simple Algebras / References

IV NOETHER'S ADDRESS TO THE 1932 INTERNATIONAL CONGRESS OF MATHEMATICIANS

    11. Hypercomplex Systems and Their Relations to Commutative Algebra and Number Theory   167

          Emmy Noether
          Notes

Publications of Emmy Noether   175

Index   179

Contributors

P. S. Alexandroff   Department of Mathematics, Moscow University, Moscow, USSR

A. Fröhlich   Department of Mathematics, King's College London, London, England

Robert Gilmer   Department of Mathematics, Florida State University, Tallahassee, Florida

Clark Kimberling   Department of Mathematics, University of Evansville, Evansville, Indiana

T. Y. Lam   Department of Mathematics, University of California, Berkeley, California

Saunders Mac Lane   Department of Mathematics, University of Chicago, Chicago, Illinois

E. J. McShane   Department of Mathematics, University of Virginia, Charlottesville, Virginia

Emmy Noether   Department of Mathematics, University of Göttingen, Göttingen, Federal Republic of Germany

Richard G. Swan   Department of Mathematics, University of Chicago, Chicago, Illinois

Olga Taussky   Department of Mathematics, California Institute of Technology, Pasadena, California

B. L. van der Waerden   Department of Mathematics, University of Leipzig, Leipzig, Germany *

---

*Current affiliation: Geschichte der Wissenschaft, Zurich, Switzerland.

EMMY NOETHER

I
BIOGRAPHY

1

Emmy Noether and Her Influence

Clark Kimberling

"She was not clay, pressed by the artistic hands of God into a harmonious form, but rather a chunk of human primary rock into which he had blown his creative breath of life." So spoke Hermann Weyl (1935) about Emmy Noether as he drew to a close his memorial address at Bryn Mawr College on April 26, 1935. Twelve days before, Emmy Noether had died following a tumor operation.

"The memory of her work in science and of her personality among her fellows will not soon pass away. She was a great mathematician, the greatest, I firmly believe, that her sex has ever produced, and a great woman." With these words, Weyl ended his compassionate but thorough memorial address. During the years since that afternoon the mathematics of Emmy Noether and the many stories about her have continued to spread throughout the world. At the centennial of her birth, it is especially appropriate to recall her life and to describe her influence in mathematics and physics.

FAMILY HISTORY

On October 14, 1806 the Prussian army fell to Napoleon at the Battle of Jena, and this event led to a wave of reforms in the status of Jews throughout what is now Germany. Also in 1806, the duchy of Baden was created, and its first constitutional edict, in 1807, recognized Judaism as a tolerated religion. The sixth such edict, in 1808, extended certain civil rights to Jews, and the ninth, the so-called Judenedict of January 13, 1809, further defined these rights and stipulated that every Jew must have an inheritable family name which satisfied certain regulations.

Accordingly, Elias Samuel, who lived in Bruchsal with his wife and nine children, received the name *Nöther,* a non-Jewish name similar to *Netter,* this latter having been used by the father of Elias Samuel. The name Nöther was passed along to a son whose given name Hertz was changed to Hermann. In 1837, Hermann joined an older brother to establish the firm of Joseph Nöther und Co., iron wholesalers. Originally located in the city of Mannheim, this firm later had branches in Düsseldorf and Berlin, until after a century of existence, it fell to anti-Jewish forces (Dick 1970). [1]

Born to Hermann and his wife Amalia Würzburger of Mannheim was Emmy's father, Max Nöther, on September 24, 1844. During his generation the spelling of his surname was changed to *Noether,* although as late as 1880, when Max was married to Ida Amalia Kaufmann, the earlier spelling was used on official papers, including the marriage certificate.

When Max was fourteen years old, he was stricken by infantile paralysis. At first he was hardly able to move at all, and after a long period of recovery which delayed his education, he remained permanently handicapped in one leg. With private tutoring, Max Noether acquired a broad education in literature and history. He studied astronomy at the observatory in his home town, Mannheim, during 1865-1866, and then studied for three semesters at the University of Heidelberg. During 1868 and 1869 he studied at the universities in Giessen and Göttingen, receiving the Ph.D. degree from Heidelberg in 1868 without a dissertation, but qualifying there in 1870 as a Privatdozent. His Habilitation paper, of considerable depth, was *Über Flächen, welche Scharen rationaler Kurven besitzen* (*Concerning Surfaces which Have Families of Rational Curves*).

In 1875 Max Noether moved to Erlangen, where the first great mathematician had been Christian von Staudt (1798-1867), a pioneer in projective geometry. World prominence had come to Erlangen through Felix Klein (1849-1925), whose Erlangen Program (Erlanger Programm in German) holds that geometries can be classified and studied according to properties which remain invariant under appropriate transformation groups. Various geometries previously studied separately were now put under one unifying theory which today still serves as a guiding principle in geometry. [See, for example, Encyclopedic Dictionary of Mathematics (1977), Kline (1972), and Klein (1893)].

Max Noether was the first of his line to earn the Ph.D. degree. Before him, his father Hermann had begun studying theology at the Lemle Moses Foundation in Mannheim, but eventually turned from academics to a career in iron wholesaling. It appears that mathematical talent came to Max through his mother's side. True, one of Max's many cousins, Ferdinand Nöther (1834-1913), was a medical doctor in Mannheim, but all the others were merchants, and the next generation included a painter, a certified engineer, and a writer. On the Kaufmann side, however, one of Max's uncles was probably gifted mathematically (Brill 1923), and among Max's descendants, one finds three mathematicians and two chemists.

The first child born to Max Noether and his wife was named Amalie Emmy Noether. Although the first name was seldom used, it is noteworthy that it belonged, with variant spelling, to both her mother and her paternal grandmother. Her mother, Ida Amalia Kaufmann, was born in 1852 in Cologne to a wealthy Jewish family whose ancestors lived on the lower Rhine. Among Ida's ten siblings, a brother, Wilhelm Kaufmann (1858-1926), became a professor at the University of Berlin and a widely recognized expert in international law and finance. Several of his books can be found in American university libraries, including a history of Germans in the American Civil War.

The other children of Max Noether were Alfred, born in 1883, Fritz in 1884, and Gustav Robert in 1889. In 1909 Fritz Noether received a doctorate in applied mathematics under Aurel Voss in Munich, and from 1911 to 1934 he worked at the Technical Institutes of Karlsruhe and Breslau. Alfred received a doctorate in chemistry in 1909 in Erlangen, where he died in 1918. Gustav Robert died in 1928.

The Noether family resided for 45 years in a multiple-family dwelling at 30-32 Nürnberg Strasse. As a child Emmy is remembered as having been not especially outstanding in any particular way. She was notably myopic, not outwardly attractive, and yet had a certain inner warmth and charm. Later, Hermann Weyl (1935) was to describe her as "warm like a loaf of bread: there irradiated from her a broad, comforting, vital warmth."

MAX NOETHER AND PAUL GORDAN

The mathematical atmosphere in which Emmy Noether grew up was determined by the work of her father and his friend and colleague Paul Gordan. No one would have guessed that Emmy would become Gordan's only doctoral student or that eventually she would assist her father in writing Paul Gordan's obituary in the *Mathematische Annalen*. Later she would assist three other authors (Castelnuovo, Enriques, and Severi 1925) in writing Max Noether's obituary in the same journal. These are among the most complete accounts of the lives and works of the two Erlangen professors

Max Noether began working at Erlangen as an ausserordentlich Professor and became an Ordinarius, or full professor, in 1888. He remained in Erlangen until his death on December 13, 1921. His most important original mathematical production came during the ten year period beginning in 1869. The Schriftenverzeichnis of his works which Emmy Noether contributed to the paper by Castelnuovo, Enriques, and Severi (1925) includes Max Noether's papers of 1869 and 1872. In these Max Noether published what is now called Noether's fundamental theorem, or his residual theorem, or his $AF + BG$ theorem. A partial statement of this important contribution to algebraic geometry will be given later in this chapter. An encyclopedia article (Berzolari 1906) gives references to proofs of this famous theorem by a dozen different authors. One proof which has been cited many times appeared in English in 1899 in *Mathematische Annalen* [See Scott (1899). [2]]

After 1879, the publications of Max Noether consisted largely of supplements and reworkings of his earlier research and that of other mathematicians. One finds in the Schriftenverzeichnis, for example, notes and comments on lectures of Riemann (Noether 1909). With Alexander von Brill (1894) he wrote a long and well-known report on the development of the theory of algebraic functions.

In his memorial address, Weyl (1935) told how the work of Max Noether fits into the history of mathematics and referred to the "scientific kinship" between father and daughter:

> ... Clebsch had introduced Riemann's ideas into the geometric theory of algebraic curves and Noether became, after Clebsch had passed away young, his executor in this matter: he succeeded in erecting the whole structure of the algebraic geometry of curves on the basis of the so-called Noether residual theorem. This line of research was taken up later on, mainly in Italy; the vein Noether struck is still a profusely gushing spring of investigations; among us, men like Lefschetz and Zariski bear witness thereto. Later on there arose, beside Riemann's transcendental and Noether's algebraic-geometric method, an arithmetical theory of algebraic functions due to Dedekind and Weber on the one side, to Hensel and Landsberg on the other. Emmy Noether stood closer to this trend of thought. A brief report on the arithmetical theory of algebraic functions that parallels the corresponding notions in the competing theories was published by her in 1920 [*sic*, actually 1919] in the *Jahresberichte der Deutschen Mathematikervereinigung*. She thus supplemented the well-known report by Brill and her father on the algebraic-

geometric theory that had appeared in 1894 in one of the first volumes of the *Jahresberichte*. Noether's residual theorem was later fitted by Emmy into her general theory of ideals in arbitrary rings. This scientific kinship of father and daughter—who became in a certain sense his successor in algebra, but stands beside him independent in her fundamental attitude and in her problems—is something extremely beautiful and gratifying. [See also Dieudonné (1972).]

Max Noether wrote a number of obituary biographies, so expertly that Weyl (1935) commented: "... such is the impression I gather from his papers and even more from the many obituary biographies ... a very intelligent, warm-hearted harmonious man of many-sided interests and sterling education." Included in the Schriftenverzeichnis of Castelnuovo, Enriques, and Severi (1925) are tributes to A. Harnack, A. Cayley, O. Hesse, J. Sylvester, F. Brioschi, S. Lie, C. von Staudt, C. Hermite, L. Cremona, G. Salmon, J. Lüroth, P. Gordan, and H. Zeuthen.

Together these writings take up some 300 pages. As summaries of important contributions to mathematics and as sources of biographical information, they have been cited many times through the years.

"Gordan was of a different stamp," Weyl wrote, "... as a mathematician, not of Noether's rank, and of an essentially different kind." Weyl (1935) then portrayed Gordan, basing his description on the obituary biography which Max Noether had written with his daughter's collaboration. Portions of the final paragraphs of this biography (Noether 1914) are translated as follows:

> For 38 years, from 1874 on, Gordan lived in Erlangen. They passed uniformly for him: daily lectures, research, and the indispensable walks either with colleagues, as earlier with Clebsch, in highly animated conversations, quite oblivious of his surroundings, or else alone in profound thought, working ideas out so completely in his head that he was able to write out the calculations at home almost without crossing out anything. Gordan had developed this remarkable mental ability long ago: Clebsch had called it "brotlosen Kunst." Even in his yearly summer vacations he carried on in the same manner, as a veritable peripatetic. Many hours were devoted to the cafe, occasionally even to chess, invariably with his cigar; evenings he liked to spend with friends or with younger members of the mathematics club; at those times he was at his liveliest, inexhaustible in stimulating discussions and chatting and in humorous remembrances. At

these gatherings, he was fond of paradoxical expressions, easily discovered by his rich imagination and powers of observation ...

... Gordan was never able to do justice to the development of fundamental concepts: even in his lectures he completely avoided all basic definitions of a conceptual nature, even that of the limit.

In the spring of 1910, Gordan stepped down from his teaching position but continued lecturing an additional four semesters, three of them along with his successor Erhard Schmidt. Gordan did not entirely stop lecturing until the semester before his passing ... With his second successor, E. Fischer, he entered a close scholarly exchange for another year. Gordan had been suffering for some time occasionally from attacks of feebleness; from such an attack that occurred in November of 1912 he did not recover: his physical and mental strength then diminished daily, until death spared him on the 21st of December 1912 ...

Noether concluded his characterization of Gordan with the simple sentence, "Er war ein Algorithmiker."

As an Algorithmiker, Gordan had been first to solve one of the major problems of invariant theory. Published in 1868, this theorem states that for each binary form $F(x_1, x_2)$ there is a *finite* set S of rational integral invariants such that every invariant of $F(x_1, x_2)$ is a rational integral function of members of S. During the decades following Gordan's long and difficult proof of this theorem, the set S, called a *complete system* or *finite basis*, was sought and found for various forms other than binary.[3] For example, Gordan himself determined complete systems for the ternary quadratic form, the ternary cubic form, and others. Not long before Gordan's retirement, his only doctoral student, the daughter of his friend Max Noether, would write her doctoral dissertation *On Complete Systems of Invariants for Ternary Biquadratic Forms* (cf. Refs 1, 2).*

CHILDHOOD AND EDUCATION (1882-1908)

"It is not quite easy," Weyl told his audience in Bryn Mawr, "to evoke before an American audience a true picture of that state of German life in which Emmy Noether grew up in Erlangen ..."

---

*Numbered references are to Noether's papers listed at the end of this book.

The great stability of burgher life was in her case accentuated by the fact that Noether (and Gordan, too) were settled at one university for so long an uninterrupted period. One may dare to add that the time of the primary proper impulses of their production was gone, though they undoubtedly continued to be productive mathematicians; in this regard, too, the atmosphere around her was certainly tinged by a quiet uniformity. Moreover, there belongs to the picture the high standing and the great solidity in the recognition of spiritual values, based on a solid education, a deep and genuine active interest in the higher achievements of intellectual culture, and on a well-developed faculty of enjoying them. There must have prevailed in the Noether home a particularly warm and companionable family life.

For eight years beginning in 1889 Emmy attended the Städtischen Höheren Töchterschule in Erlangen, where she studied under very nearly the same curriculum as is found in the comparable German schools of today. At home, like most daughters, Emmy cleaned and dusted around the house and learned to cook. She took piano lessons, but unlike her mother, who was accomplished at piano, Emmy did not excel in music. On the other hand, she was particularly fond of dancing. In school she became proficient in French and English, and during five days in April 1900, she took the Bavarian State Examination for teachers of French and English. Her final grade in each language was "very good," and only for the test in practical classroom conduct was her grade merely "satisfactory." Her success with these examinations entitled her to teach foreign languages at female educational institutions. However, before entering into such a career, she decided to study at the university.

This decision encountered resistance from many sides, for in Germany in 1900 very few women attended universities. Women had been allowed to enroll at universities in France in 1861, England in 1878, and Italy in 1885. However, in Germany in 1900, professors frequently refused to grant permission to women to attend lectures, and only very rarely was a woman allowed to take university examinations. The eminent historian Heinrich von Treitschke became a spokesman for traditionalists who genuinely thought that the presence of women in universities would undermine the fundamental purposes of academic life. Following is a sample, as translated in Evans (1976, p. 17) of von Treitschke's point of view:

> Many sensible men these days are talking about surrendering our universities to the invasion of women, and thereby falsifying their

entire character. This is a shameful display of moral weakness. They are only giving way to the noisy demands of the press. The intellectual weakness of their position is unbelievable ... The universities are surely more than mere institutions for teaching science and scholarship. The small universities offer the students a comradeship which in the freedom of its nature is of inestimable value for the building of a young man's character ...

At the University of Erlangen, home of the Noether family, the Academic Senate in 1898 went so far as to declare that the admission of women students would "overthrow all academic order."

Nevertheless, in 1900, Emmy Noether was able to get permission to attend lectures. The registry at the University of Erlangen shows that during the winter semester, only two of the 986 students were females. Emmy continued to attend lectures until 1902, and on July 14, 1903 she took and passed the matura examinations at the royal Realgymnasium in nearby Nürnberg.

During the winter semester of 1903-1904, Emmy Noether was registered as a student at the University of Göttingen, where she attended the lectures of the astronomer Karl Schwartzchild (1873-1916) and the mathematicians Hermann Minkowski (1864-1909), Otto Blumenthal (1876-1944), Felix Klein (1849-1925), and David Hilbert (1862-1943).

After the one semester in Göttingen, Emmy Noether returned to Erlangen, for it had just become legally possible for female students to matriculate there and to take examinations with the same rights as male students. Matriculating as student number 468 on October 24, 1904, Emmy Noether listed only mathematics as her course of study.

Three years later she finished her doctoral dissertation under Paul Gordon at Erlangen. Entitled *Über die Bildung des Formensystems der ternären biquadratischen Form* (*On Complete Systems of Invariants for Ternary Biquadratic Forms*), the dissertation was officially registered with the date July 2, 1908 as number 202. A note based on the dissertation, published in 1907 [1], was Emmy Noether's first publication. Her second, in 1908 [2], is a reproduction of the dissertation in its entirety.

These two publications and others for the next decade represent the first of three epochs of Emmy Noether's scientific production as perceived by Hermann Weyl. The years up to 1919 he called the period of relative dependence, first upon the Gordan style of problems and methods and then upon those of Hilbert, the transition taking place under the influence of the Erlangen mathematician Ernst Fischer (1875-1954). The striking difference between the mathematics of Gordan and that of Hilbert was portrayed by

Gordan himself in response to Hilbert's famous solution of the so-called Gordan problem in the theory of invariants. Far removed from the formal, algorithmic, constructive approach of Gordan, Hilbert in 1888 obtained a proof "by contradiction" of the existence of a finite basis for certain invariants. This was the problem Gordan had labored on for years. When he finished reading Hilbert's existence proof, Gordan exclaimed, "Das ist nicht Mathematik; das ist Theologie."

Hilbert dealt the theory of invariants another blow in 1892, after which he summarized the history of the theory: The first of three periods, the *naive*, began with Cayley and Sylvester in England. "In the drawing up of the simplest invariant concepts and in the elegant applications to the solutions of equations of the first degree, they experienced the immediate joy of first discovery" (Reid 1970). The second period Hilbert called the *formal*, as represented by Clebsch and Gordan. The final period, the *critical*, Hilbert claimed as his own, and it was characterized largely through his theorems of 1888 and 1892. Algebraic invariant theory was finished, except for its applications; a later mathematician wrote that "the breath went out from the whole theory" (Reid 1970; see also Fisher 1967, Fogarty 1969, and Sloane 1977).

It is therefore understandable that after Emmy Noether had made the transition from computational methods to the Hilbert approach, she referred to her dissertation with the words "Mist," "Rechnerei," and "Formelngestrüpp"—a jungle of formulas. The dissertation concludes with a voluminous tabulation of 331 ternary quartic covariant forms, a fact which led to a somewhat disparaging final sentence in the review of the work in *Jahrbuch über die Fortschritte der Mathematik*.

During the years 1908-1915 Emmy Noether worked without compensation at the Mathematical Institute in Erlangen. Aside from carrying out her own research, she occasionally substituted for her father at his lectures, as he was becoming increasingly hindered with advancing age. In 1908 she was elected to membership in the Italian organization Circolo matematico di Palermo, and in 1909 she joined the Deutschen Mathematiker-Vereinigung. With considerable zeal she attended meetings and gave presentations. She was able especially at the many less formal sessions "Mathematik zu reden" (to talk mathematics).

THE FIRST EPOCH (1908-1919)

During the first epoch of Emmy Noether's scientific production, that is, the period of dependence of which Weyl spoke, Gordan retired. His successor,

Erhard Schmidt (1876-1959), had little impact in Erlangen. Then came Ernst Fischer (1875-1959),[4] who in Weyl's judgment asserted a more penetrating influence on Emmy Noether than had Gordan. Frequently the two attended the mathematics seminar, and soon afterward she would send Fischer postcards which extended the mathematical discussions. According to Dick (1970), when one reads this correspondence (which ranges from 1911 to 1929, peaking in 1915), one gets the impression that Emmy Noether often began writing immediately after talking with Fischer, continuing the train of thought as though there had been no interruption.

In her manuscript *Körper und Systeme rationaler Funktionen* [6], which she finished in May 1914, Emmy Noether noted that the work was motivated by conversations with Fischer. B. L. van der Waerden (1935) writes that this was Emmy Noether's most important paper during her years in Erlangen.

Again, in a 1919 curriculum vitae, Emmy Noether wrote that Ernst Fischer had had a deciding influence in bringing her to a more abstract way of thinking mathematically. As Fischer was a specialist in invariant theory and elimination theory (the study of solutions of systems of equations, not necessarily linear), and as he was a proponent of Hilbert's style of thought, Emmy Noether became more and more adept at the Hilbert approach to invariant theory. By 1915 she was so accomplished in this area that Klein and Hilbert invited her to the University of Göttingen. Emmy wrote from Göttingen in November 1915 to Fischer: "Invariantentheorie ist hier Trumpf; sogar der Physiker Hertz studiert Gordan-Kerschensteiner; Hilbert will nächste Woch über seine Einsteinschen Differentialinvarianten vortragen, und da müssen die Göttinger doch etwas können." (*Invariant theory is trump here;... Hilbert is planning to lecture next week on his Einsteinian differential invariants, and to understand that, the Göttingen people must certainly know something!*)

It should be remarked that the term *invariant theory* as used here means *differential* invariant theory, as contrasted to *algebraic* invariant theory, although the latter, especially in the hands of Emmy Noether, was a powerful tool for the former.

Before leaving Erlangen, Emmy Noether served as dissertation adviser for Hans Falckenberg in 1911 and for Fritz Seidelmann in 1916. Regarding the latter, Emmy had already moved to Göttingen but was able to advise Seidelmann during vacations, which she spent working at her parents' home in Erlangen. Seidelmann, like his adviser Emmy Noether nine years earlier, passed his doctoral oral examination summa cum laude. An excerpt of Seidelmann's dissertation is included in the conclusion of Emmy Noether's paper *Gleichungen mit vorgeschriebener Gruppe* [11].

Returning to Göttingen after her mother's death in May 1915, Emmy Noether was at first occupied with the questions which led to Ref. 11, that is, to equations with prescribed group, as first discussed by Dedekind and then presented to her by Landau. By 1917 her interests had become centered on the connections between invariant theory and Einstein's relativity theory. Weyl (1935) wrote:

> Hilbert at that time was over head and ear in the general theory of relativity, and for Klein, too, the theory of relativity and its connections with his old ideas of the Erlangen program brought the last flareup of his mathematical interests and mathematical production. The second volume of his history of mathematics in the nineteenth century bears witness thereto. To both Hilbert and Klein, Emmy was welcome as she was able to help them with her invariant-theoretic knowledge. For two of the most significant sides of the general theory of relativity theory she gave at that time the genuine and universal mathematical formulation.

In a letter dated May 24, 1918, Einstein wrote to Hilbert:

> Gestern erhielt ich von Frl. Noether eine sehr interessante Arbeit über Invariantenbildung. Es imponiert mir, dass man diese Dinge von so allgemeinem Standpunkt übersehen kann. Es hätte den Göttinger Feldgrauen nichts geschadet, wenn sie zu Frl. Noether in die Schule geschickt worden waeren. Sie scheint ihr Handwerk zu verstehen! [5]

Already, Emmy Noether was recognized for the extreme generality and abstractness of approach which would eventually be seen as her most distinguishing characteristic.

Emmy Noether had worked unpaid in Erlangen. At Göttingen, Hilbert and Klein tried to get something better for her. It was on November 9, 1915 that she gave a lecture entitled *Über ganze transzendente Zahlen*, which was to fulfill a requirement for Habilitation. If the faculty found the work acceptable, she would (under ordinary circumstances) receive the title Privatdozent and the accompanying privilege, in the form of a venia legendi, of lecturing without pay under university sponsorship. Emmy Noether wrote to Fischer after her lecture: "Das hat sich sogar der hiesige Geograph angehört, für den es ein bisschen sehr abstrakt war; die Fakultät will sich in ihrer Sitzung von den Mathematikern keine Katze in Sack verkaufen lassen." (*Even the local geographer attended, though he found it a bit abstract; the faculty wants no pig in a poke sold them in their session by the mathematicians.*)

As it turned out, the faculty had nothing sold to them at all, since the promotion for which Emmy Noether was being considered could not proceed because, according to a 1908 Privatdozentenverordnung, only males could be candidates for Habilitation. A separate petition for compensation for Emmy Noether was also rejected.

In his memorial address Weyl revealed that it was the philologists and historians in the Philosophical Faculty, which included the mathematicians, who blocked Hilbert's efforts on Emmy Noether's behalf. They argued: Must the soldiers, returning from the deprivations and discipline of war, now find themselves being lectured at the feet of a woman? At a faculty meeting, Hilbert countered: "I do not see that the sex of the candidate is an argument against her admission as Privatdozent. After all, we are a university, not a bathing establishment."

"Probably," Weyl wrote, "he provoked the adversaries even more by that remark." Nevertheless, Emmy Noether was permitted to lecture under Hilbert's name. Indeed, the official lecture schedule for Winter Semester 1916-1917 included *Mathematisch-physikalischer Seminar. Invariantentheorie: Prof. Hilbert mit Unterstützung von Frl. Dr. E. Noether, Montags 4-6, gratis.* Other such announcements show that Noether regularly assisted Hilbert, not only in the seminar and practicums but also in the main lectures, through the summer of 1919.

During the years of World War I, and especially during 1917-1918, Emmy Noether worked on differential invariants. On August 22, 1917, she wrote to Fischer, "Die Differentialinvarianten, deren Beweise ich im Frühjahr erst unvollständig hatte, habe ich nun tatsachlich auf ein Äquivalenzproblem linearer Scharen zurückgeführt; es wäre schön, wenn es sich Ihrer Theorie einordnen liesse ... (*The differential invariants, whose proof I had only partially finished in the spring, I have now reduced to an equivalent problem about linear families. It would be nice if this could be tied in with your theory ...*)

The material on differential invariants [12, 13] was written up and published in 1918. By 1919 the war had ended, and the change in the political system, as well as other sweeping changes, had greatly improved the legal status of women in general and Emmy Noether's chances for Habilitation in particular.

Accordingly, on May 21, 1919 an application was submitted to the faculty; on May 28 the required colloquium was held, and on June 4, Emmy Noether presented her Habilitation lecture to the mathematicians at Göttingen.

Among the papers for Emmy Noether's Habilitation is found, in her own handwriting, a summary of her research which up to that time had been published or was ready for publication. A translation of this curriculum vitae follows:

> My dissertation and a later paper ... belong to the theory of formal invariants, as was natural for me as a student of Gordan. The lengthiest paper, *Fields and Systems of Ratonal Functions*, concerns questions about general bases; it completely solves the problem of rational representation and contributes to the solution of other finiteness problems. An application of these results is contained in *The Finiteness Theorem for Invariants of Finite Groups*, which provides by a completely elementary finiteness proof an actual basis. To this line of investigation also belongs the paper *Algebraic Equations with Prescribed Group*, which yields a contribution to the construction of such equations over an arbitrary domain of rationality ... The paper *Integral Representation of Invariants* proves valid a conjecture of D. Hilbert and at the same time gives a purely conceptual proof for the series development of invariant theory, based on the equivalence of families of linear forms and partially fashioned on ideas of E. Fischer. This work then led to further work by E. Fischer ... With these wholly algebraic works belong two additional works which are not yet published: *A Proof of Finiteness for Integral Binary Invariants* ... and an investigation with W. Schmeidler of noncommutative one-sided modules ... In this context belongs also work concerned with questions in algebra and the theory of modules modulo g and with the question of *Alternatives with Nonlinear Systems of Equations* ... The longer work *The Most General Ranges of Completely Transcendental Numbers* uses, along with algebraic and number-theoretic techniques, some abstract set theory ... In this same direction is the paper *Functional Equations and Isomorphic Mapping*, which yields the most general isomorphic mapping of an arbitrary abstractly defined field.
>
> Finally, there are two works on differential invariants and variation problems. These resulted from my assistance to Klein and Hilbert in their work on Einsteinian general theory of relativity ... The second of these, *Invariant Variation Problems*, which is my Habilitation paper, treats arbitrary continuous Lie groups, finite or infinite, and draws conclusions from a special case of

invariance relative to such a group. These general results contain as special cases some known results concerning proper integrals from mechanics, stability theorems, and certain dependencies among field equations arising in the theory of relativity, while, on the other hand, the converses of these theorems are also given.

The last few sentences of the curriculum vitae refer to the "genuine and universal mathematical foundation," as Weyl described it in 1935. Concerning this contribution to physics. Peter G. Bergmann later wrote:[6]

> Noether's theorem, so-called, forms one of the corner stones of work in general relativity as well as in certain aspects of elementary particles physics. The idea is, briefly, that to every invariance or symmetry property of the laws of nature (or of a proposed theory) there corresponds a conservation law, and vice versa. Accordingly, if a physical quantity is known to satisfy a conservation law (known as a "good quantum number" in quantum physics), the theorist attempts to construct a theory with appropriate symmetry properties. Conversely, if a theory is known to possess certain symmetries, then this fact alone entails the existence of certain integrals of the dynamical equations.
>
> General relativity is characterized by the principle of general covariance, according to which the laws of nature are invariant with respect to arbitrary curvilinear coordinate transformations that satisfy minimal conditions of continuity and differentiability. A discussion of the consequences in terms of Noether's theorem (whether explicitly quoted as such or not) would have to include all of the work on ponderomotive laws, *inter alia*.
>
> Goldstein's text *Classical Mechanics* contains a treatment of Noether's theorem on pp. 47 ff., without, however, calling it by that (or any other) name. J. L. Anderson's book *Principles of Relativity Physics* (Academic Press, 1967) explicitly refers to Noether's theorem on p. 92.

Emmy Noether's Habilitationsschrift [13], in which Noether's theorem was first published, has been translated into Russian (Polak 1959) and into English (Tavel 1971). The work continues to be of great importance in contemporary physics, as indicated by Science Citation Index, Logan (1977), Houtappel, Van Dam, and Wigner (1965), and Ibragimov (1972).

Emmy Noether and Her Influence

The distinguished Russian topologist P. S. Alexandroff praised Emmy Noether's contributions to algebraic and differential invariants in a speech (1936) in her memory which he gave, as president of the Moscow Mathematical Society, at a meeting of the Society on September 5, 1935: "These works ... would themselves have secured her reputation as a first-rate mathematician, and they represent no less a contribution to mathematical science than the notable researches of S. V. Kovaleskaya."

Weyl too mentioned Sonya Kovaleskaya (1850-1891) in his memorial address on Emmy Noether, but as much for sharp contrast in personality as for comparison in mathematics. Kovaleskaya, after studying in Berlin with Karl Weierstrass (1815-1897) for four years, received her Ph.D. summa cum laude from the University of Göttingen in 1874. Remembered also as an author of books which portray Russia during 1860-1880, Kovaleskaya was Professor of Mathematics at the University of Stockholm, a member of the editorial board of *Acta Mathematica*, and a Corresponding (i.e., honorary) Member of the Russian Academy of Sciences (Polubarinova-Kochiva 1957).

Alexandroff (1936) in his Moscow speech continued, "But when we speak of Emmy Noether as a mathematician ... we mean not so much these early works, but instead, the period beginning about 1920 when she struck the way into a new kind of algebra, becoming the leader of ... begriffliche Mathematik."

THE SECOND EPOCH (1920-1926)

The second of the three distinct epochs into which Weyl classified Emmy Noether's scientific production he described as "the investigations grouped around the general theory of ideals, 1920-26." He wrote:

> Her development into that great independent master whom we admire today was relatively slow. Such a later maturing is a rare phenomenon in mathematics ... Sophus Lie, like Emmy Noether, is one of the few great exceptions. Not until 1920, thirteen years after her [receiving the Ph.D.], there appeared in the *Mathematische Zeitschrift* that paper of hers written with Schmeidler, *Über Moduln in nichtkommutativen Bereichen, insbesondere aus Differential- und Differenzenausdrücken* [17], which seems to mark the decisive turning point. It is here for the first time that the Emmy Noether appears whom we all know, and who changed the face of algebra by her work ... I suppose that Schmeidler gave as much as he received in this cooperation.

Max Noether. (Courtesy of Archiv, Friedrich-Alexander-Universität, Erlangen.)

Paul Gordan. *(Courtesy of Archiv, Friedrich-Alexander-Universität, Erlangen.)*

Ernst Fischer. (Courtesy of Ursula Fischer.)

B. L. van der Waerden, June 1931. (Courtesy of B. L. van der Waerden.)

Paul S. Alexandroff and Heinz Hopf, Zurich, 1931. (Courtesy of Otto Neugebauer.)

Helmut Hasse. (Courtesy of Jutte Kneser, née Hasse.)

Emmy Noether, standing at the Göttingen Railroad Station, about to depart for the United States, 1933. (Courtesy of Otto Neugebauer.)

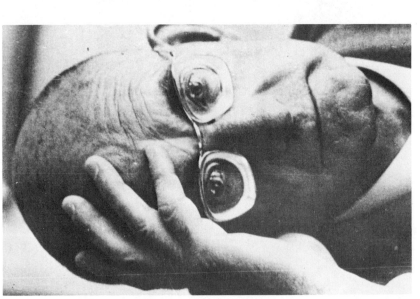

Paul Alexandroff. (Courtesy of P. S. Alexandroff and A. Maltsev.)

Group picture, Nikolausberg, near Göttingen, 1932. Left to right: Ernst Witt, Paul Bernays, Helene Weyl, Hermann Weyl, Joachim Weyl, Emil Artin, Emmy Noether, Ernst Knauf, unidentified person, Chiungtze Tsen, Erna Bannow (later Mrs. E. Witt).

Mathematisches Institut, Göttingen. (Courtesy of Ann Gray.)

Göttingen, Stegemühlenweg 51, 18.8.33.
(bis 25.8 Dierhagen (Mecklenburg)).

To the President
 of Bryn Mawr College.

My dear Dr. Park,

Ich danke Ihnen und dem College, vor allem dem Department of Mathematics, vielmals für das mir überraschend angebotene Stipendium. Es war mir eine große Freude und ich werde es mit Vergnügen annehmen.

Ich muß allerdings um eine Verschiebung um ein Jahr bitten, auf das akademische Jahr 1934/35. Aber ich hoffe und vermute daß das keine allzu großen Schwierigkeiten machen wird. Ich habe nämlich für diesen Winter, für den Weihnachts-Oster-Term, schon eine Einladung nach Oxford angenommen, in das Somerville College, zu Gastvorlesungen an der Universität die auch den Mathematikern der andern Colleges zugänglich sein werden. Bis zur Beendigung dieser Vorlesungen ist auch das akademische Jahr fast beendet.

Es ist mir persönlich ein sehr angenehmes Gefühl schon für zwei Jahre im voraus meine Pläne machen zu können; ich hoffe sicher daß sich alles gut regeln wird.

Ihre sehr ergebene
Emmy Noether.

Den Katalog habe ich mit Interesse gelesen.

Emmy Noether's letter, 8/18/33, to President Park. (Courtesy of Archive, Bryn Mawr College.)

In later years, however, Emmy Noether frequently acted as the true originator; she was most generous in sharing her ideas with others.

In N. Bourbaki, *Elemènts d'histoire des mathématiques* (1960), it is pointed out that the familiar concepts of *right ideal* and *left ideal* originate with the 1920 Noether-Schmeidler paper. The next year Emmy Noether's *Idealtheorie in Ringbereichen* [19] appeared in *Mathematische Annalen*; this was later called "revolutionary" by I. Kaplansky (1970) because of its influence on the development of the general theory of ideals. This paper demonstrated the fundamental importance and naturalness of chain conditions on ideals, which had previously been studied by Dedekind, Lasker, and others. Largely as the result of this paper, rings in which the ascending chain condition holds are called *noetherian rings*.

The years 1920-1922 were a period of transition in the administration of mathematics at Göttingen. Klein, ready to retire, was successful in getting Richard Courant appointed as his successor. A clear accounting of the transition, one which at the same time reveals university life as Emmy Noether knew it, is given by Constance Reid in her books *Hilbert* (1970) and *Courant* (1976). In *Courant*, one reads that "In Göttingen in 1920 there was no mathematics department . . . no budget [except for the library] . . . no building—in fact, the professors did not even have offices." Two years later, under Courant, "mathematics and the natural sciences were formally separated from the other specialties of the Philosophical Faculty," and permission was granted for the use of stationery bearing the letterhead Mathematisches Institut der Universität.

On April 6, 1922 the Prussian Minister for Science, Art, and Public Education sent Emmy Noether an explanation of the designation nicht beamteter ausserordentlicher Professor, a title which was now hers, "mit dem Bermerken," the words were exactly, "dass diese Bezeichnung eine Änderung ihrer Rechtsstellung nicht zur Folge hat. Insbesondere bleiben die aus Ihrer Stellung als Privatdozentin zu Ihrer Fakultät sich ergebenden Verhältnisse unberührt; auch ist damit die Übertragung einer beamteten Eigenschaft nicht verbunden." Thus, as an "unofficial associate professor," Emmy Noether had a title without accompanying responsibilities and salary. After some urging of the authorities by Courant, she was also granted a Lehrauftrag für Algebra on April 22, 1923. This provided a small but regular remuneration for teaching.

Now able to guide students officially up to the doctoral examinations, Emmy Noether once again became a dissertation adviser and doctoral examiner. Having already overseen the work of Falckenberg and Seidelmann at Erlangen, she served as adviser for nine more who studied for their doctorates at Göttingen.

First there was Grete Hermann, who received her doctorate on February 25, 1925. Next there were Heinrich Grell and Rudolf Hölzer. The latter died of tuberculosis at 23, after his dissertation was finished but before actually receiving his degree. Heinrich Grell received his doctorate on July 14, 1926, and has since, on a number of occasions, expressed his indebtedness to Emmy Noether. In his own words, Emmy Noether "guided his inclinations into the right path," especially by her introducing him to the work of Dedekind.

In the fall of 1924, a 22-year-old Dutchman who had finished the university course at Amsterdam came to Göttingen on the recommendation of L. E. J. Brouwer (1881-1966). This was B. L. van der Waerden, who, it is said, was allowed to study mathematics as a child only after his father realized that young Bartel was inventing trigonometry and using notations of his own instead of the "right" ones.

Alexandroff (1936) wrote that van der Waerden, once he had arrived in Göttingen, readily mastered the theories of Emmy Noether, enhanced them with findings of his own, and like no one else, promulgated her ideas. "A course in the general theory of ideals, given by van der Waerden in 1927 in Göttingen, was enormously successful." This was only a beginning of his expository treatment of material which originated with Emmy Noether. His famous book *Moderne Algebra* (1931) was based on the lectures of E. Noether and E. Artin.

Garrett Birkhoff (1973) states that "van der Waerden's *Moderne Algebra* precipitated a new revolution" in the history of algebra.

> As I have indicated, both the axiomatic approach and much of the content of "modern" algebra dates back to before 1914. However, even in 1929, its concepts and methods were still considered to have marginal interest as compared with those of analysis in most universities, including Harvard. By exhibiting their mathematical and philosophical unity, and by showing their power as developed by Emmy Noether and her younger colleagues (most notably E. Artin, R. Brauer, and H. Hasse), van der Waerden made "modern algebra" suddenly seem central in mathematics. It is not too much to say that the freshness and enthusiasm of his exposition electrified the mathematical world — especially mathematicians under 30 like myself.

In December 1971 Professor Birkhoff wrote to van der Waerden, asking for information on the main sources of *Moderne Algebra*. This led to a very interesting article by van der Waerden (1975) in *Historia Mathematica,* in which he writes in the same clear manner of exposition which 45 years earlier had spread the mathematics of Emmy Noether:

> In the beginning of our century, many people felt that the theory of invariants was a mighty tool in algebraic geometry. According to Felix Klein's Erlanger Programm, every branch of geometry is concerned with those properties of geometrical objects that are invariant under a certain group. However, when I studied the fundamental papers of Max Noether, the "Father of Algebraic Geometry" and the father of Emmy Noether, and work of the great Italian geometers, notably of Severi, I soon discovered that the real difficulties of algebraic geometry cannot be overcome by calculating invariants and covariants . . .
>
> Another problem that worried me very much was the generalization to n dimensions of Max Noether's "fundamental theorem on algebraic functions." Noether's theorem specified the conditions under which a given polynomial $F(x,y)$ can be written as a linear combination of two given polynomials f and ϕ with polynomial coefficients A and B: $F = Af + B\phi$. More generally, one can ask under what conditions a polynomial $F(x_1, \ldots, x_n)$ can be written as a linear combination of given polynomials $f_1, \ldots, f_r$, with polynomial coefficients: $F = A_1 f_1 + \cdots + A_r f_r$, or in modern terminology, under what conditions is F contained in the ideal generated by $f_1, \ldots, f_r$. From the papers of Max Noether I knew that this question is of considerable importance in algebraic geometry, and I succeeded in solving it in a few special cases. I did not know then that Lasker and Macaulay had obtained much more general results . . .
>
> It seems that Hilbert was the first to realize that an n-dimensional generalization of Noether's theorem would be desirable. Emmanuel Lasker, the chess champion, who took his Ph.D. degree under Hilbert's guidance in 1905, was the first to solve this problem. He proved that, quite generally, every polynomial ideal $(f_1, \ldots, f_r)$ is an intersection of primary ideals.
>
> In her 1921 paper *Idealtheorie in Ringbereichen* [19] . . . , Emmy Noether generalized Lasker's theorem to arbitrary commutative rings satisfying an "ascending chain condition" (Teilerkettensatz). Chapter 12 of my book *Allgemeine Ideal-*

theorie der kommutativen Ringe[7] is based on this paper of Emmy Noether.

Van der Waerden explains that the ascending chain condition, which is the defining property of noetherian rings, is not a very stringent condition. If a ring satisfies stronger assumptions, then more can be proved about the ring. Under stronger assumptions than "noetherian," Emmy Noether proved a ring-theoretic Fundamental Theorem of Arithmetic. Specifically, in her paper *Abstrakter Aufbau der Idealtheorie in algebraischen Zahl- und Functionenkörpern* [31], she gave five axioms for what are now called *Dedekind rings*. Her theorem states that every ideal in a Dedekind ring is a product of prime ideals. "In these rings," van der Waerden writes, "Dedekind's theory of ideals in algebraic number fields and fields of algebraic functions of one variable is valid." Dedekind rings, specifically those which are integral domains, are presented in Chapter 14 of the first edition of *Moderne Algebra*.

Emmy Noether greatly admired the work of Richard Dedekind (1831-1916). This is evident not only in her own mathematics, such as the formulation of the five axioms under which Dedekind's theory of ideals holds, but also in her teaching. Weyl wrote, "Of her predecessors in algebra and number theory, Dedekind was most closely related to her. For him she felt a deep veneration. She expected her students to read Dedekind's appendices to Dirichlet's 'Zahlentheorie' not only in one, but in all editions." And at the beginning of van der Waerden's paper on the sources of *Moderne Algebra*, one finds prominently displayed:

Motto:
Es steht alles schon bei Dedekind. [8]
Emmy Noether

Concerning the first few chapters of *Moderne Algebra*, van der Waerden states that he learned group theory "mainly from Emmy Noether's course *Gruppentheorie und hyperkomplexe Zahlen* (winter 1924/25) and from oral discussions with Artin and Schreier in Hamburg." In Chapter 6, the "wording and proof of the two isomorphy theorems of § 40 are due to Emmy Noether. The same is true for § 42." (All references are to the first edition, in German.) Van der Waerden refers to Emmy Noether's paper *Eliminationstheorie und allgemeine Idealtheorie* [24] when he writes,"... one may say that the whole Chapter 11 was closely connected with Emmy Noether's work on elimination theory and on my own work on the foundation of algebraic geometry."

Van der Waerden (1948) also writes in his historical survey that Emmy Noether taught him "that algebraic geometry ought to be based on Steinitz' algebraic theory of fields, and on Dedekind's arithmetical theory of algebraic functions and ideals."

"I saw at once that she was right," van der Waerden continues, "for I had learnt to see the importance of fields and ideals by my previous work. I was enthused by her foundation of the general theory of ideals, based upon Hilbert's basis theorem, and by her theory of 'ganz-abgeschlossene Ringe' [integrally closed rings]."

Before van der Waerden arrived in Göttingen in 1924, he had studied under L. E. F. Brouwer in Holland. In his historical note on topology in Holland, Alexandroff (1969) describes Emmy Noether's visit to Blaricum during Christmas vacation 1925-1926. "She was at that time deeply absorbed in her group-theoretic lecture at Göttingen." In contact with Brouwer and Alexandroff, Emmy Noether, as early as the time of her visit, perceived applications of group theory to the foundations of combinatorial topology. "In particular, the numerical invariants—the Betti numbers and torsion coefficients—should be replaced by homology groups." Alexandroff (1969) continues with the remark that several topologists, including Lefschetz, were at first skeptical about this, but that this profound insight by Emmy Noether eventually found general acceptance among topologists. [9]

The summer following Emmy Noether's visit in Holland, Alexandroff came to Göttingen. With his new friend and colleague Heinz Hopf (1894-1971), he "belonged to the circle of mathematicians associated with Courant and Emmy Noether, an unforgettable group with its musical evenings and boating excursions at Courant's place and in company with him, and with its 'algebraisch-topologischen Spaziergängen' led by Emmy Noether." Frequently the mathematical discussions were carried on at the swimming pool at the university. Located on the Leine River, the swimming pool was under the direction of Fritz Klie, "a thoroughly striking personality" among the Göttingen university people. "Many a mathematical—and not only mathematical—discussion took place 'beim Klie': along the flowing—not always clean—in fact after a rain really rather brown—water of the Leine, or in the sun, or in the shade (which was shared with the gnats) of the lovely trees of the Klie Swimming Pool. Many a mathematical idea was born there" (Alexandroff 1969).

The Klie Swimming Pool was "for men only" but two women used it anyway. These were Emmy Noether and Mrs. Nina Courant: the two "took advantage of their exclusive rights daily, regardless of the weather."

It is well known (e.g., Reid 1970) that Brouwer and Hilbert maintained a highly strained relationship with each other—one in which Weyl was caught in the middle. In the summer of 1926, Brouwer visited Göttingen and immediately joined the circle of Courant and Emmy Noether.

The mathematical atmosphere in this circle was so lively and the human side so warm, that all the ice in human relationships was obliged to melt away. Accordingly, plans were made to bring about a reconciliation between Brouwer and Hilbert. An evening meal at Emmy Noether's place was selected. In the cozy attic room which served as the study, the living room, and the dining room in Emmy Noether's apartment, Brouwer, Hilbert, Courant, and Landau were seated, along with some younger mathematicians including Hopf and myself. It was my task to direct the conversation toward the reconciliation.

Now, it is a well-known fact that the best way to bring two people back together is to center the focus on some third party whom the first two parties are eager to criticize. Following this principle I brought the conversation around to the noted functional theoretician from Luckenwald—the one who had achieved fame for his work on the theory of uniformity. The success of this undertaking surely exceeded our boldest expectations, for in a short time Hilbert and Brouwer were pressing each other in a high-spirited exchange of opinion, and drawing closer and closer to a common point of view regarding the third party. All the while they nodded to each other in a friendlier and friendlier fashion. At last they united in a mutual toast to each other (Alexandroff 1969).

The reconciliation lasted for the rest of Brouwer's visit, but later the two great men reverted to their former ways.

That winter, Hermann Weyl was in Göttingen as a visiting professor, and he conducted a lecture on representations of continuous groups which was attended by Emmy Noether. "I have a vivid recollection . . . ," he wrote (1935), "for just at that time the hypercomplex number systems and their representations had caught her interest and I remember many discussions when I walked home after the lectures, with her and von Neumann, who was in Göttingen as a Rockefeller Fellow, through the cold, dirty, rain-wet streets of Göttingen."

THE THIRD EPOCH (1927-1935)

The third and final epoch of Emmy Noether's production recognized in Weyl's memorial address concerns noncommutative algebras, especially as represented by linear transformations and as applied to commutative number fields. Insight into her interests in 1927 is provided by van der Waerden (1975):

When I came to Göttingen, I took Emmy Noether's course *Gruppentheorie und hyperkomplexe Zahlen* in 1924/25. One of the main subjects in this course was Maclagan Wedderburn's theory of algebras over arbitrary fields. The same subject was treated, in a much improved form, in her course under the same title in 1927/28, in which also a quite new treatment of representations of groups and algebras was given. I took notes of the latter course, and these notes formed the basis of Emmy Noether's publication in *Mathematische Zeitschrift 30* (1929), p. 641. The Chapter 16 (*Theorie der hyperkomplexen Grössen*) and 17 (*Darstellungstheorie der Gruppen und hyperkomplexen Grössen*) [in *Moderne Algebra*] are almost entirely due to Emmy Noether. Only §127, on the representations of the symmetric groups S_n, comes from an oral communication by John von Neumann, as stated in a footnote.

Elsewhere in his paper (1975), van der Waerden points out that Section 108 in Chapter 15 (*Linear Algebra*) was also drawn from the 1929 paper in *Mathematische Zeitschrift*.

Especially after the formal establishment of the Mathematische Institut, Göttingen became, in the opinion of many, the most important center of mathematical activity in the world. Felix Klein's dream that Göttingen would become the "Mecca of mathematics," having been entrusted to Richard Courant as Director of the Institut, had come to fulfillment within a few years after Klein's death in 1925. By the early 1930s the circle of algebraists around Emmy Noether had gained recognition as the most active part of the Institut (Weyl 1935).

An especially vivid account of the "glow" at Göttingen and its sudden extinguishing was written by Constance Reid (1976) in her biography of Courant. Under Courant's leadership, the Institut recovered dramatically from the ravages of war and inflation. Students abounded in greater numbers than ever before. Scholars from around the world began visiting again, including a number of leading Russian mathematicians. P. S. Alexandroff and P. S. Urysohn came first in 1923. In following years, more or less under Alexandroff's initiative, the visiting Russians included L. S. Pontryagin, A. O. Gelfond, L. F. Schnirelman, L. A. Lusternik, O. J. Schmidt, and A. N. Kolmogoroff.

Courant arranged for Alexandroff, who spoke excellent German, to give regular lectures each year from 1926 to 1930. These were very well attended. Although Alexandroff's work was primarily in topology, he nevertheless became a close friend of Emmy Noether. His was the wit which dubbed her "der Noether," the German word *der* serving more as an adjective than as the definite article which precedes nouns of masculine gender. When the Russians "began to go around Göttingen in their shirtsleeves—a startling departure from proper dress for students—the style was christened 'the Noether-guard uniform' " (Reid 1976).

These and other anecdotal remembrances [e.g., in Dick (1970), Polyá (1973), Reid (1970), Reid (1976)] of Emmy Noether's outward appearance, her heaviness, and the volume of her voice—"It was often not easy for one to get the floor in competition with her"—these remembrances reveal the real-life presence of Emmy Noether among those who knew her in Göttingen. Even today it is recalled, sometimes with affection and sometimes not, that she thought little about what she should wear, what she should eat, and so on. Her intentions could hardly be further removed from the effects which her appearance had, especially on those who did not know her. In particular, there was the handkerchief. She kept it tucked under her blouse. While lecturing she had a way of jerking it out and thrusting it back in, frequently and energetically, and this was very noticeable to her audience. Too, during the years before she began keeping her hair cropped short, she wore it up, and little by little during the excitement of lecturing, it would fall out of place. "The Graces did not stand at her cradle," Weyl (1935) wrote; and a widely circulated wall chart (IBM 1966) states matter-of-factly that "she was fat, rough, and loud, but so kind that all who knew her loved her."

During the summer of 1926, Heinz Hopf was in Göttingen, and he and Alexandroff soon became close friends. Emmy Noether tried to arrange for Alexandroff to have a permanent professorship in Germany but was unable to get anywhere with this idea. She was successful, however, in prevailing upon Hermann Weyl to recommend Alexandroff and Hopf for Rockefeller scholarships. The two topologists were able to spend the year 1927-1928 together with the leading topologists Oswald Veblen (1880-1960) and Solomon Lefschetz (1884-1972) at the Institute for Advanced Study. [10]

Beginning in 1929 the mathematicians at Göttingen were housed in the new Mathematische Institut on Bunsenstrasse.

> The new building—a three-level T-shaped structure—provided everything the mathematicians had ever needed or wanted in their physical surroundings. Courant always gave Neugebauer the credit for its planning. The basement contained such requirements as a bicycle room, a book bindery, and a room for refreshments. The double stairs of the main entrance led to a spacious lobby, which today contains a bust of Hilbert and is known to students and faculty as "the Hilbert space." On the main floor there were

two large auditoriums, *Maximum* and *Minimum*, and four other rooms of various sizes—all equipped for lectures in applications as well as in theory—a mechanical drawing room, a room for the meetings of the Praktikum and of the mathematische Gesellschaft, smaller meeting rooms, offices, individual workrooms, consulting rooms. On the top floor were the spacious and well-planned quarters of the Lesezimmer, still the heart of the mathematical life of Göttingen (Reid 1976).

Among the offices in the new building, called *"Kabuffs"* in Göttingen jargon, one of the larger was assigned to Emmy Noether. Her work during these years was rooted in a large body of results by Molien, Frobenius, Dickson, Wedderburn, and others. As earlier, in her work with ideals, Emmy Noether's methods in noncommutative algebras had a unifying effect. Always they were of a highly conceptual nature, avoiding calculative tools where possible in favor of automorphisms and other transformations which preserve essential aspects of structure. "In intense cooperation with Hasse and with Brauer," Weyl wrote (1935), "she investigated the structure of noncommutative algebras and applied the theory by means of her *verschränktes Produkt* (cross product) to the ordinary commutative number fields and their arithmetics."

From the epoch of 1927 to 1935, Emmy Noether's two major publications were *Hyperkomplexe Grössen und Darstellungstheorie* [35] in 1929 and *Nichtkommutative Algebren* [41] in 1933. Also there were three lesser papers on norm residues and the principal genus theorem. Hasse (1932) published an account of Emmy Noether's verschränktes Produkt in his work on cyclic algebras. Noether, Hasse, and Brauer published in 1932 a paper entitled *Beweis eines Hauptsatzes in der Theorie der Algebren* [39]. This paper, which Weyl praised as an enduring "high mark in the history of algebra," proved that every simple algebra over an ordinary algebraic number field is a cyclic algebra according to the definition by Dickson.[11]

During the winter of 1928-1929 Emmy Noether was a visiting professor at the University of Moscow. In his memorial address P. S. Alexandroff (1936) describes her interest in and ready adaptation to the "Moscow ways" and gives many examples of her influence on mathematicians in the Soviet Union.

The year 1932 was special for Emmy Noether. During this, her last full year in Germany, she received with Artin the Alfred Ackermann-Teubner Memorial Prize for advancement of mathematical science. The award of 500 reichmarks was given "for her total scientific production." This was one form of visible recognition which was given her in 1932. Another was the major address presented by her at a general session at the International Mathematical Congress held in Zurich in September.

Among the official delegates to the Congress, Hermann Weyl represented the Deutsches Mathematiker-Vereinigung, Edmund Landau the Gesellschaft der Wissenschaften zu Göttingen, Richard Courant the University of Göttingen, Helmut Hasse the University of Marburg, and Wolfgang Krull and Otto Haupt the University of Erlangen. There were approximately 800 people present for the Congress. Of the 420 participants, Emmy Noether was one of the 21 who gave major addresses. This international recognition was certainly a high point in her career, but there remains today the realization that Emmy Noether was not recognized as she should have been during her lifetime. In particular, she was never elected to membership in the Göttinger Gesellschaft der Wissenschaften, nor was she ever made an ordentliche Professor (full professor). In his memorial address, P. S. Alexandroff sharply rebukes "high-ranking Prussian academic bureaucracy" for its failure to recognize Emmy Noether properly.

For her fiftieth birthday celebration, the algebraists at Göttingen gathered, and Hasse (1933) dedicated to Emmy Noether a paper in which he derives a reciprocity law without using commutativity, thereby justifying an idea of hers that certain aspects of noncommutative algebra are characterized by a greater regularity than those in the commutative case. Emmy Noether was very happy about this paper and its being dedicated to her.

With Robert Fricke (1861-1930) and Oystein Ore (1899-1968), Emmy Noether edited the collected mathematical works of Richard Dedekind (1930-1932). It was planned that Fricke would contribute a biography of Dedekind from personal remembrance, but he died before this work was completed. Relatively few of the many editorial comments were written by Fricke; most of this work was done by Noether and Ore. Noether's many comments show her deep insight into Dedekind's mathematics and to some extent these also represent her own contributions to the further development.

Together with Jean Cavaillès, who perished as a supporter of the resistance to the Nazis in his native France, Emmy Noether (1937) edited the correspondence of Georg Cantor (1845-1918) and Richard Dedekind. Although ready for publication in 1933, the volume was not published until 1937. In the foreword, in French, Cavaillès recalls the enjoyable days he spent in Göttingen, and he lauds Emmy Noether for her good cheer, hospitality, and the intense intellectual radiance which was hers.

The 1960s and 1970s have seen a renewed interest in the work of Richard Dedekind. For example, Kenneth O. May (1972) wrote in a review of a Russian publication: "... the author makes a well documented case for upgrading the traditional estimate of [Dedekind's] role." Dedekind lectured in group theory in Göttingen a full 13 years before Jordan, it has traditionally been thought, launched the subject; in quite a different direction, Dedekind stated the Peano axioms many years before Peano; and so on.[12] Many years ago, Emmy Noether recognized the full stature of this man.

In the 1932 elections, the Nazis won more seats in the Reichstag than did any of the rival parties. As leader of the Nazis, Adolf Hitler became Chancellor of the Reich on January 30, 1933. At the University of Göttingen, pro-Nazi students suddenly became very active. Brown shirts and swastikas became a common sight in the lecture halls. The SA (Sturmabteilung, or "storm troops," the Nazi militia), were given "freedom of the streets."

On March 21, Hitler announced the beginning of the Third Reich. He then passed the Enabling Act, under which he issued decrees autonomously. In the revolution which followed, Germany was converted from a democratic republic into a totalitarian state with Hitler as dictator. Nazi anti-Semitic policies were put into practice. Jews were systematically disqualified from taking part in German national and cultural life.

The 1925 census showed that the 564,379 Jews living in Germany comprised about 0.9% of the total population. Fully one-third of these lived in Berlin. Thus the Jews in German universities, because of their number, were in a particularly conspicuous position.

At the University of Göttingen, the directors of three of the four institutes of physics and mathematics were Jewish, and all three—James Franck, Max Born, and Richard Courant—resigned or were put on leave before May 1933.

The German Students Association at Göttingen started a campaign "Against the un-German Spirit," according to which Jews were able to think only Jewish, and their works should be printed only in Hebrew—or perhaps translated into German by a real German. When a Jew wrote in German it was all false. Professors and students should be screened "according to their guarantee of thinking in the German spirit" (Beyerchen 1977, p. 16). One very active member of the pro-Nazi group was Werner Weber, who had taken his doctorate under Emmy Noether, had been Edmund Landau's assistant, and had been a Privatdozent in Göttingen since 1931.

Emmy Noether received the following notification from a representative of the Prussian Minister for Sciences, Art, and Public Education, under the designation U I Nr. 17277: "Auf Grund des §3 des Berufsbeamtentums vom 7. April 1933 entziehe ich Ihnen hiermit die Lehrbefugnis an der Universität Göttingen. (*On the basis of paragraph 3 of the Civil Service Code of April 7, 1933, I hereby withdraw from you the right to teach at the*

University of Göttingen.) On April 26, the newspaper, the *Göttinger Tageblatt* announced under the ominous subtitle "More will follow" that six professors had been placed on leave, and three of the names given were Born, Courant, and Noether. See Dresden (1942), Pinl and Furtmüller (1973), and Ringer (1969).

The dream which Klein, Hilbert, Courant, and others had brought to reality over the decades was struck at its roots. Hermann Weyl, who had come to Göttingen in 1930 to fill the position vacated by Hilbert's retirement, took over as acting director of the Mathematische Institut after Courant was forced out. "A stormy time of struggle like this one we spent in Göttingen in the summer of 1933," Weyl (1935) wrote, "draws people closely together; thus I have a particularly vivid recollection of these months. Emmy Noether— her courage, her frankness, her unconcern about her own fate, her conciliatory spirit—was in the midst of all the hatred and meanness, despair and sorrow surrounding us, a moral solace."

A particularly disturbing experience during the last half of 1933 involved Werner Weber. As the leader of a group of pro-Nazi students, he commanded a boycott of a lecture of Edmund Landau. "In front of his lecture hall," Courant wrote to Abraham Flexner, "were some seventy students, partly in S.S. uniforms, but inside not a soul. Every student who wanted to enter was prevented from doing so by [Weber]." Landau was soon informed by a "scientifically gifted ... but completely muddled and notoriously crazy" young man [probably Oswald Teichmüller] that "Aryan students want Aryan mathematics and not Jewish mathematics ..." These excerpts from Courant's letter are quoted by Reid (1976).

During those tumultuous days, Emmy Noether kept mathematics before all else. She had been well known for calling her students and colleagues together sometimes even during holidays when, officially, the university was closed. So it was on one occasion in 1933 when the Noether group gathered in her apartment to discuss the preparations of a lecture by Hasse on class field theory. This was a very private sort of gathering—one intended only for a select group of persons who were interested in the subject. To the surprise of others attending, one student[13] of whom Emmy Noether was particularly fond took part in the meeting wearing an SA uniform. It is remembered that Emmy Noether was not visibly disturbed. In fact, when the incident was recalled a few weeks later in a conversation with van der Waerden, she laughed about it. Later, even Hasse became a "fervent supporter of the Nazis" (Feit 1979; Segal 1980).

"Her heart knew no malice;" Weyl (1935) wrote, "she did not believe in evil—indeed it never entered her mind that it could play a role among men. This was never more forcefully apparent to me than in the last stormy summer, that of 1933, which we spent together in Göttingen."

IN AMERICA (1933-1935)

> *It shall not be forgotten what America did during these last two stressful years for Emmy Noether and for German science in general.*
>
> Hermann Weyl

At first, after the venia legendi—the right to teach at the University of Göttingen—was taken away, Emmy Noether continued to meet informally with students and colleagues, but it was at the same time clear that she would soon have to move elsewhere, at least temporarily. As early as July 11, 1933, President Marion Edwards Park of Bryn Mawr College wrote to the Rockefeller Foundation in New York:

> I have received a letter from E. R. Murrow of the Institute of International Education informing me that the sum of $2000 has been granted to Bryn Mawr College by the Emergency Committee to Aid Displaced German Scholars[14] for the support of a German scholar to be chosen by Bryn Mawr. The Committee assumes that the scholar chosen should be offered a salary for $4000 for the year and suggests that I write to the Rockefeller Foundation asking whether the Foundation would see fit to make an additional grant of $2000 to Bryn Mawr for a year in order to complete the amount of the salary ... Earlier in the spring Professor Anna Pell Wheeler of our Department of Mathematics was approached by Professor Lefschetz of Princeton with the suggestion that Dr. Emmy Nothe [*sic*] of Göttingen whose resignation from the University had been demanded should be invited to Bryn Mawr College for 1933-34.

A few days later, on the other side of the Atlantic, the Principal of Somerville College of Oxford University contacted the Paris headquarters of the Rockefeller Foundation: "I have received intimation both from the University Registry here and from the Academic Assistance Council that the Rockefeller Foundation is generously disposed ... German scholars ... [and] I write at once to say that I am in negociation with Frau Professor Emmy Noether of Göttingen University ..."

There followed a number of letters, radiograms, and telegrams involving the Paris and New York offices of the Rockefeller Foundation and President Park. The situation became so complicated that on September 18, 1933, Warren Weaver, then Director of the Division of Natural Sciences of the International Education Board of the Rockefeller Foundation, wrote a four-page letter to President Park. A less official and closer look at the matter is recorded in a log submitted to the Paris office by Professor H. M. Miller, dated September 25, 1933:

Frau Prof. EMMY NOETHER. In view of the uncertainty whether N. is to go to Oxford or to Bryn Mawr, HMM called on Miss Helen Darbyshire, Principal of Somerville College ... D. understands the R.F. position, and said that after N. has come to Oxford she (D.) would hope to make an arrangement with Cambridge, whereby N. would spend the last term of the year there. With reference to the Bryn Mawr offer, Miss Darbyshire, who some years ago spent one year at Wellesley as a visiting professor, expressed the opinion that Bryn Mawr would not be a suitable place for Prof. N. This statement was made flatly, but HMM did not press D. for her reasons for thinking so.

Miss D. is a classicist and knows nothing about the field of mathematics. She had only recently returned from vacation and was not at all sure what action her College could take in finding funds to supplement the £24 now available for N. She is a very pleasant person, but somewhat at a loss to know how to proceed in the practical matter of extending aid to Prof. N.

After a good bit of subsequent correspondence, Warren Weaver wrote to President Park his assurance that "the Foundation may be counted on to contribute the sum of $2000 toward the salary of Prof. Noether at Bryn Mawr during the academic year 1934-1935, in addition to the Foundation's contribution toward Prof. Noether's salary for the year 1933-34," and on October 2, Weaver received word from President Park:

DOCTOR EMMY NOETHER HAS CABLED TODAY ACCEPTING APPOINTMENT AT BRYN MAWR FOR THIS YEAR.

The news spread. Notes in the Rockefeller Archive Center include one from C. S. Gibson to Lauder Jones: "Max Born said the other day that Frau Noether is going to Bryn Mawr ..." and Professor Miller's log for November 16 states: "Prof. Polya said that he had heard that Prof. NOETHER had sailed for the U.S. and went on to say that with the exception of Prof. Hilbert, Miss Noether was unquestionably the most important teacher in Germany, especially keen in attracting and developing young research workers ... "

The earliest record of any connection between Emmy Noether and Bryn Mawr College may very well be the letter of July 11, 1933. Before 1933, perhaps Emmy Noether had never had any more interest in Bryn Mawr College than in any other high-ranking college. Nevertheless, in retrospect one must recall that Charlotte Angas Scott, the first mathematician at Bryn Mawr and department head from 1885 to 1924, had published (Scott

1899) a proof of Max Noether's fundamental theorem in *Mathematische Annalen*. Scott's successor as department head, Anna Johnson Pell Wheeler (1883-1966), nearly the same age as Emmy Noether, had a mathematical background which is well worth recounting here.

Having been born in Iowa of Swedish immigrants, Anna was graduated from the University of South Dakota in 1903. She received her master's degree from the University of Iowa in 1904, a second master's degree from Radcliffe in 1905, and then moved to Göttingen. There she attended lectures by Hilbert, Klein, and Minkowski, and became especially interested in integral equations.

In July 1907, Anna was married to Alexander Pell. For this occasion, Pell, at some risk to his life, had journeyed from Vermillion, South Dakota, to Göttingen. (Pell, whose real name was Sergei Degaev, was an escaped Russian double agent, a fact not generally known until long after his death.) The two returned to Vermillion, but soon Anna was back in Göttingen again, where she intended to write a doctoral dissertation under Hilbert. Unsuccessful in this, she returned to the United States. The Pells moved to Chicago, and in 1910 Anna received her Ph.D. under E. H. Moore. In 1918 she started teaching at Bryn Mawr College, and in 1924 she succeeded Charlotte A. Scott as chairman of the mathematics department (Grinstein and Campbell 1978).

Another figure in the history of Bryn Mawr College whose influence must have been felt by Emmy Noether was Martha Carey Thomas (1857-1935), the daughter of one of the first trustees of Johns Hopkins University. With an 1877 degree from Cornell, Miss Thomas moved to Germany. After three years' study at Leipzig she transferred to the University of Zurich, since it was almost impossible for a woman to receive a Ph.D. in Germany at that time. With a dissertation in English philology, she received her Ph.D. at Zurich the same year Emmy Noether was born. Three years later Bryn Mawr College was founded. M. Carey Thomas was appointed dean and professor of English, and from 1894 to 1922 she was President of the College.

The enormous role which M. Carey Thomas played in the development of the character of Bryn Mawr College is indicated by Veysey (1971): "... Germany, and her own scholarly diligence, existed for her primarily as tokens of woman's capability. She was transforming herself into an example ... Germany connoted rigorous, painful excellence; therefore it was necessary and useful in her struggle to prove to Americans that women could achieve the same intellectual distinction as men ..." [See also Thomas (1979).]

Four years before M. Carey Thomas stepped down from the presidency, Marion Edwards Park (1875-1960) finished her Ph.D. at Bryn Mawr College.

A scholar in classics who had taught and served as a dean at two other colleges, Miss Park became President of Bryn Mawr College in 1922.

It was she who wrote to Warren Weaver on November 28, 1933: "I am venturing to ask you whether by any lucky chance you can come down to Bryn Mawr in December and see Dr. Noether in action! We are inviting all the mathematicians of the neighborhood to hear her speak at half past four on the afternoon of Friday the 15th ... A dozen or more of the faculty members from Princeton, Swarthmore and the University will meet her at dinner at the Deanery, the alumnae house of the college."

President Park continued, "Dr. Noether is settled in what I hope are comfortable quarters, her English proves to be entirely usable and she is already much enjoyed by everyone who has come in contact with her. She has begun some work informally with the students and the department expects to make a more formal arrangement at the beginning of the next semester."

On March 28, 1934 President Park reported to the Rockefeller Foundation that Bryn Mawr would be

> ... appointing for next year, in addition to the usual fellow and scholar in mathematics a second fellow and scholar, with stipends of $860 and $400 respectively, to be called the Emmy Noether Fellow and Scholar. The Emmy Noether Fellowship has been awarded to Miss Carolyn Shover,[15] an advanced student of mathematics who already holds her Ph.D. from Ohio State University and has taught Mathematics for four years at Connecticut College. The Emmy Noether Scholarship was awarded to Miss Marie Weiss, a Ph.D. of Stanford University who held a National Research Fellowship for two years at the University of Chicago. Unfortunately we shall be able to give next year only one of the five fellowships which the college has usually offered for foreign women, but that one fellowship is to be held by a student in mathematics, Dr. Olga Taussky, a Czechoslovakian who received her doctor's degree from the University of Vienna in 1930, and who was for several years assistant at the University of Göttingen.[16] The nomination of Dr. Taussky was made by Dr. Noether herself.

Warren Weaver kept a diary in which, during May 1934, he wrote a particularly insightful paragraph about Emmy Noether:

> N. seems reasonably satisfied, in her characteristically philosophical fashion, with the situation at Bryn Mawr. She says

there are two girls who are really interesting to her, one of them she thinks will do an interesting piece of work. She comes to Princeton one day a week so that she does not lack for scientific contacts. She speaks with great enthusiasm and some stubbornness concerning Max Zorn, feeling that he should receive special consideration on acount of his ability. It is clear, however, that N's interest in Z. is primarily based on the fact that he works along lines which are closely related to those of N. herself.[17]

When Emmy Noether returned to Germany during the summer of 1934, she found that things were not at all the way she had left them. At first she visited the Artins in Hamburg and then went on to Göttingen. There she found very few of her former acquaintances. As a visiting "foreign scholar" she was allowed to use the mathematics library.

Natascha Artin (now Brunswick) recalled[18] that her husband had a "method" for talking about mathematics with Emmy.

They would go for walks, and he would ask her a question, and she would talk *very*, very fast. He knew he couldn't keep up with her, so he would let her talk for about half an hour and then say "Emmy, but I didn't understand a word; could you please tell me again." And she would start again. But in the meantime they would walk very fast, and she would get a little slower and go more slowly through it again. The second time he would say, "Emmy, I haven't understood it yet." On the third rendition he would understand what she was talking about. By that time, you see, she was so tired that her speed would slow down. She was so amazingly lively!

Not long after Emmy Noether returned to Bryn Mawr, she joined the American Mathematical Society, and a number of people became involved in the process of "absorbing" her into a suitable academic environment. Part of the problem, again, was her salary.

In order to secure continued support for Emmy Noether's residence at Bryn Mawr College, Jacob Billikopf, through Arnold Dresden, who was Chairman of the Department of Mathematics and Astronomy at Swarthmore College, solicited letters from Solomon Lefschetz, Norbert Wiener, and George D. Birkhoff.

Lefschetz' letter, dated December 31, 1934, begins as follows:

Dear Mr. Billikopf:
   Professor Dresden has requested that I write to you regarding Professor Emmy Noether's place in the mathematical world.

This will not take me very long; she is the holder of a front rank seat in every sense of the word. As the leader of the modern algebra school, she developed in recent Germany the only school worthy of note in the sense, not only of isolated work but of very distinguished group scientific work. In fact, it is no exaggeration to say that without exception all the better young German mathematicians are her pupils. Were it not for her race, her sex and her liberal political opinions (they are mild) she would have held a first rate professorship in Germany and we would have no occasion to concern ourselves with her. She is the outstanding refugee German mathematician brought to these shores and if nothing is done for her, it will be a true scandal.[19]

On January 2, 1935, Wiener wrote to Billikopf:

... Miss Noether is a great personality; the greatest woman mathematician who has ever lived; and the greatest woman scientist of any sort now living, and a scholar at least on the plane of Madame Curie. Leaving all questions of sex aside, she is one of the ten or twelve leading mathematicians of the present generation in the entire world and has founded what is certain to be the most important close-knit group of mathematicians in Germany—the Modern School of Algebraists. Even after she was deprived of her position in Germany on account of her sex, race and liberal attitude, numbers of students (men as well as women) continued to meet at her rooms for mathematical instruction. Of all the cases of German refugees, whether in this country or elsewhere, that of Miss Noether is without doubt the first to be considered.[19]

George D. Birkhoff's letter, dated January 5, 1935, follows:

Dear Mr. Billikopf:

At the suggestion of Professor Arnold Dresden of Swarthmore College, I want to write you in connection with Miss Emmy Noether, now at Bryn Mawr College. It will, indeed, be very helpful if you can succeed in securing funds to make possible the payment of a moderate salary to Miss Noether. She is generally regarded as one of the leaders of modern Algebraic Theory. Within the last ten or fifteen years she and her students in Germany have led the way much of the time. It is not too much to say that, since Sonia Kavalevski, she is the only woman mathematician of high absolute rank. Thus it is an opportunity for us to have her in this country and I hope very much that you will succeed in your efforts. Her continued presence at Bryn Mawr is sure to be a stimulus to everyone interested in modern Algebra in this country.

As far as the desirable arrangements to be made in her case are concerned I find myself in general agreement with my colleague Professor Wiener. I might mention here the fact that at least when I was last in Germany her salary there was not large, and also that as far as undergraduate work is concerned, she will be probably of no use at Bryn Mawr.[19]

These letters Billikopf sent to Alfred E. Cohn, who answered on January 8 that "the Emergency Committee [in Aid of Displaced German Scholars] will ... make a determined effort to continue to supply grants. Need I say that your offer of $750 to $1000 cheerfully placed at the disposal of the Emergency Committee for Professor Noether will be cheerfully accepted. The proper procedure is for Bryn Mawr College to request a renewal of the existing grants from the Emergency Committee and the Rockefeller Foundation."

There were difficulties, however, as the Rockefeller Foundation had determined that "assistance will be restricted to those cases where there is a definite plan for the absorption, through an eventually permanent post, of the deposed scholar in question."

Nevertheless, another insightful entry in Warren Weaver's diary, dated March 20, 1935, shows that the Rockefeller Foundation was, after further consideration, prepared to contribute to Emmy Noether's salary for two more years:

President M. E. Park (Bryn Mawr) presents case of Dr. Emmy Noether. It has become clear that N. cannot possibly assume ordinary academic duties in this country. She has no interest in undergraduate teaching, has not made very much progress with the language, and is entirely devoted to her research interests. On the other hand the Bryn Mawr authorities like and admire her, and would very much like to have her around for a further period. An interested friend of N's has obtained personal pledges amounting to $1,700. The Princeton Institute for Advanced Study will devote $1,500 toward her support in return for a seminar which N. conducts on her weekly visit to Princeton. P. agrees that $3,000 a year is sufficient for N. Thus an additional $2,800 is necessary to provide for two further years. There is no hope whatsoever of absorption at Bryn Mawr, but there appears to be a fair chance for absorption at the Princeton Institute, this being an ideal disposition of the case.

WW agrees to give consideration, making it clear that N's case falls entirely outside of the standard formula, and that we are justified only on the grounds of her unusual eminence and upon the further grounds that it is pointless to insist upon absorption in the case of a person who is incapable of being absorbed.

In later conversation, *MM* [Max Mason; see Weaver (1970)] agrees that the case warrants special consideration. WW discusses on the telephone with *Murrow*, who agrees to present the case to his committee for half of the necessary $2,800, the RF to furnish the other half.

Weaver confirmed these arrangements in a letter to President Park, dated April 8, and she wrote him on April 12 that the various sources for Dr. Noether's salary of $3000 for 1935-1936 and $3000 for 1936-1937 were now established.

On April 10, however, Emmy Noether had undergone surgery. Earlier, the doctors had thought "she was not a good operative risk."

Accordingly, we had her go to the hospital two days before operation in order to let her rest and to have laboratory studies made that would determine her general physical condition.

Because of her high blood pressure and her general physical make-up, we felt that spinal anesthesia rather than inhalation anesthesia would be indicated in her case.

At operation the pelvic tumor was found to be a large ovarian cyst the size of a large cantaloupe. The cyst was removed. There were also two small fibroid tumors in the uterus which were not removed because they were not producing symptoms and it was deemed advisable not to prolong the operation unnecessarily. In accordance with Dr. Noether's request, the appendix was removed.

Dr. Noether had a perfectly normal convalescence during the first three post-operative days ...

During the early morning of the fourth post-operative day she developed a circulatory collapse from which she seemed to rally under treatment. At noon on that day she suddenly lapsed from consciousness ... From that point, Dr. Noether failed rapidly in spite of every effort to save her.[20]

"The patient died in the early afternoon, the temperature before death reaching the extraordinary figure of 109 degrees," wrote another of the physicians, "... it is not easy to say what had occurred in Dr. Noether. It is possible that there was some form of unusual and virulent infection, which struck the base of the brain where the heat centers are supposed to be located."[21]

This was on a Sunday. The next morning, Warren Weaver read about the tragedy in his newspaper and sent a note of sympathy to President Park. On the campus, friends of Emmy Noether gathered at President Park's home on Wednesday afternoon. There, in a simple service, Hermann Weyl and Richard Brauer, having come over from Princeton, and Anna Pell Wheeler and Olga Taussky spoke briefly.

On Thursday, Marguerite Lehr spoke in the Bryn Mawr College Chapel. She recalled the excitement with which the faculty had awaited Emmy Noether's arrival in 1933. "When she came, all ... barriers were suddenly nonexistent, swept away by the amazing vitality of the woman whose fame as the inspiration of countless young workers had reached America long before she did ..."

Miss Lehr continued,

> Professor Brauer, in speaking yesterday of Miss Noether's powerful influence professionally and personally among the young scholars who surrounded her in Göttingen, said that they were called the Noether family, and that when she had to leave Göttingen, she dreamed of building again somewhere what was destroyed then. We realize now with pride and thankfulness that we saw the beginning of a new "Noether family" here. To Miss Noether her work was as inevitable and natural as breathing, a background for living taken for granted; but that work was only the core of her relation to students. She lived with them and for them in a perfectly unselfconscious way. She looked on the world with direct friendliness and unfeigned interest, and she wanted them to do the same. She loved to walk, and many a Saturday with five or six students she tramped the roads with a fine disregard for bad weather. Mathematical meetings at the University of Pennsylvania, at Princeton, at New York, began to watch for the little group, slowly growing, which always brought something of the freshness and buoyancy of its leader.
>
> Outside of the academic circle, Miss Noether continually delighted her American friends by the avidity with which she gathered information about her American environment. She was

proud of the fact that she spoke English from the very first; she wanted to know how things were done in America, whether it were giving a tea or taking a Ph.D., and she attacked each subject with the disarming candor and vigorous attention which won every one who knew her.

From Princeton, Abraham Flexner, Director of the Institute for Advanced Study, wrote to President Park, in a letter dated April 25, 1935:

Her death has shed a deep gloom over us all, but it ought to make you and Mrs. Wheeler happy to know that a few weeks ago she remarked to Professor Veblen that the last year and a half had been the very happiest in her whole life, for she was appreciated in Bryn Mawr and Princeton as she had never been appreciated in her own country.

Within a few months, obituary tributes appeared in several languages— Einstein (1935), Barinaga (1935), van der Waerden (1935), Berra (1935), Weyl (1935), Kořínek (1935), Alexandroff (1936)—and in the years since then the significance of Emmy Noether as a scientist and as a person has come to be more fully recognized and understood. Among the many scientific and biographical writings since 1960 are Chee (1975), Dick (1970), Dubreil-Jacotin (1961), Humphreys (1978), Kimberling (1972), Mac Lane (to appear), Noether (1976), Osofsky (1974), Smith (1976), Tavel (1971), and Wussing and Arnold (1978).

Today on the campus of Bryn Mawr College, Goodhart Hall stands where it has since Hermann Weyl delivered his memorial address there in 1935. A short distance to one side of Goodhart stands the President's House, and quite close, on the other side of Goodhart, stands Rockefeller. These and other buildings surround the well-known structure at the heart of the campus: the M. Carey Thomas Library, named for the early President of Bryn Mawr College who, as a pioneer in the higher education of women in America, brought rigorous German standards of excellence to her college and country. Around the cloisters of the M. Carey Thomas Library there is a brick walk under whose pavement are buried the ashes of President Thomas. Buried there also are the ashes of the great mathematician, Emmy Noether.

STUDENTS, COLLEAGUES, AND INFLUENCE

Recalling those students and colleagues who were closest to Emmy Noether, Weyl (1935) wrote of "the Noether boys," including, at one time or another, W. Krull, H. Grell, G. Köthe, M. Deuring, E. Witt, C. Tsen, K. Shoda, and

J. Levitski. "She lived in close communion with her pupils; she loved them, and took interest in their personal affairs." In Göttingen they were regarded as something of a family, and a rather "noisy and stormy family" at that. Norbert Wiener (1956) wrote that, while on the train to the 1932 Mathematical Congress in Zurich, he with his own family encountered their friend Emmy Noether, "probably the best woman mathematician there has ever been ..., looking like an energetic and very near-sighted washerwoman ... and her many students flocked around her like a clutch of ducklings about a kind, motherly hen."[22]

Five of Emmy Noether's doctoral students have already been mentioned (Falckenberg, Seidelmann, Hermann, Grell, and Hölzer). Chronologically, the next was Werner Weber (1906-194?), who received his doctorate in 1929 with a dissertation on *Idealtheoretische Deutung der Darstellbarkeit beliebiger natürlicher Zahlen durch quadratische Formen*. Weber contributed minor portions to the editing of Dedekind's works (1930-1932), e.g., vol. II, p. 384, and his assistance is acknowledged in the Preface to the first edition of *Moderne Algebra* (van der Waerden 1931).

Jakob Levitski (1904-1956), born in Ukraine, had emigrated to Palestine with his parents. Coming to Göttingen from Tel Aviv, he had very little money, but was recognized especially by Emmy Noether for his mathematical talent. She tried diligently to get an assistantship for him. She wrote (Dick 1970) that he was "ausserordentlich tüchtig und sympathisch, ..., nichts von unangenehm jüdisch" (*extraordinarily industrious and likeable, ... with no trace of an unpleasant Jewish quality*). Eventually, Levitski was able to go to Yale University as a Sterling Scholar. Beginning in 1931, he lectured at the Hebrew University in Jerusalem. In 1954 he was awarded the Israel Prize for Exact Sciences. The Levitski radical in associative ring theory is named after him (Amitsur 1974).

Another of the "Noether boys," Max Deuring, was considered by Emmy Noether to be the most promising student at Göttingen. Deuring wrote his dissertation under Noether in 1930. Three years later she wanted him to be her successor at Göttingen, after she had emigrated to the United States. After appointments in Marburg and Hamburg, he taught at the University of Göttingen for many years, beginning in 1950.

Among Deuring's most notable contributions have been those in pseudovaluations, zeta functions, and the solution of Riemann's modulus problem in the parametrization of algebraic curves of genus 1 in arbitrary characteristic. His monograph (1968), often cited in the literature, acknowledges the influence of Emmy Noether and includes a survey of her work in hypercomplex number systems.

Still another "Noether boy" was Hans Fitting, who received his Ph.D. in 1931. Emmy Noether secured a scholarship for him from the Notgemeinschaft der Deutschen Wissenschaften, and with this support he worked in Göttingen and Leipzig prior to his Habilitation at Königsberg. The impact of Fitting's work in algebra is manifest in the widely used concepts *Fitting group* and *Fitting radical*. Unfortunately, Fitting died of a bone disease when only 32 years old.

After Fitting, Emmy Noether's next doctoral candidate to receive his degree was Ernst Witt. She wrote (Dick 1970), regarding Witt, "Er hat auf einmal angefangen zu arbeiten und nicht nur zu vereinfachen." (*He started to work all of a sudden, and not merely to simplify.*) Completing his studies at the time of the Nazi takeover, Witt had all but finished his doctoral work, working in the range of interests of Noether, Hasse, and F. K. Schmidt at that time. As Emmy Noether's position at Göttingen was terminated in April 1933, Witt's doctoral work was assigned to Gustav Herglotz (1881-1953), and Witt received his Ph.D. in July 1933.

During his career Ernst Witt made a number of contributions which now bear his name: *Witt vectors* in connection with valuation rings, *Witt's theorem* and *Witt group* in the study of quadratic forms over a general field, *Witt's theorem* on ϵ-trace forms, the *Hasse-Witt matrix,* and the *Poincaré-Birkhoff-Witt theorem* in universal enveloping algebras (in the theory of Lie algebras). Witt also contributed substantially to class field theory, Mathieu groups, and the word problem (in the study of free groups).

On December 6, 1933, Emmy Noether's student Chiungtze Tsen received his Ph.D. with a dissertation entitled *Algebren über Funktionenkörpern.* Tsen returned to China, where he was a professor at Chekiang University in Hangchow during 1935-37. His paper (1936) appears in the first issue of the Journal of the Chinese Mathematical Society. In the summer of 1937 he became a professor at Pei-yang University in Tientsien. Then, during the war with Japan, Tsen moved, along with the university, to Sian in Shensi province. In 1938 he moved again to Sikang in Szechwan province. There some time in 1939 or 1940, he perished from a disease.[23] Although political conditions and premature death cut short Tsen's career, he is nevertheless remembered for Tsen's Theorem: A normal simple algebra over a field K of algebraic functions of one variable over an algebraically closed field is a total matrix algebra over K (Encyclopedic Dictionary of Mathematics 1977).

The last doctoral student of Emmy Noether at Göttingen was Otto Schilling. Having studied extensively also the work of Hasse, Schilling took his degree at Marburg after Emmy Noether's emigration. In 1935, he too emigrated to the United States. His most notable publication has been *The*

Theory of Valuations (1950), a culmination of work begun by Kurt Hensel (1861-1941) and developed by J. Kürschák, to a remarkable extent by A. Ostrowski, and further by Artin and Krull.

Emmy Noether once exclaimed that "only foreigners" came to her. The fact is, among her visiting colleagues and students were P. S. Alexandroff from Russia, Chiungtze Tsen from China, Kenjiro Shoda from Japan, G. Köthe from Austria, J. Herbrand from France, J. Levitski from Israel, and B. L. van der Waerden from Holland. Of particular importance in carrying the influence of Emmy Noether to Japan was Kenjiro Shoda.

Having studied group representation theory under Teiji Takagi (1875-1960) at Tokyo Imperial University and then under Issai Schur at the University of Berlin, Kenjiro Shoda in 1927 moved to Göttingen. Some 26 years earlier Takagi had journeyed to Göttingen to study under Hilbert. Now his brilliant student Shoda became one of the Göttingen "Noether boys." In the words of Hirosi Nagao (1978), "This particular year seems to mark the most significant period in his mathematical growth. There, near Noether, he witnessed the remarkable process of creation of great mathematical ideas ... , and youthful Shoda buried himself in enthusiastic pursuit of mathematics [among] the many young, able mathematicians who had come from all over the world to Göttingen then, attracted to Emmy Noether." This impetus led to Shoda's many notable contributions to the theory of algebras, among which was the Brauer-Shoda Galois theory for simple algebras.

In 1932 Shoda published 抽象代数学 (*Abstract Algebra*). For its influence on many Japanese algebraists, this book may be compared with van der Waerden's *Moderne Algebra* (1931). Among the many who learned algebra from Shoda's books were Tadashi Nakayama, Keizo Asano, Masaru Osima, and Goro Azumaya. Shoda went on to play an important role in the founding of the Mathematical Society of Japan and of the Osaka Journal of Mathematics.

After publishing 41 research papers and seven books, Kenjiro Shoda became president of Osaka University, a position which he held from 1935 to 1961.

In 1928 Emmy Noether wrote to Hasse that there was only one French mathematician whose research was closely related to algebra as treated by the Noether circle. This was A. Châtelet (1883-1960). Beginning in the 1930s, however, several young Frenchmen made their way to Göttingen to study mathematics. Claude Chevalley entered into a lively sharing with Hasse and Noether, and his publication (1933) on norm residues was, for example, inspired by Noether's lecture during the winter of 1929-1930.

The state of mathematics in France at that time was summarized by Jean A. Dieudonné (1970) in an address before the Roumanian Institute of Mathematics in 1968.

Masters like Picard, Montel, Borel, Hadamard, Denjoy, Lebesgue, etc., were living and still extremely active, but these mathematicians were nearly fifty years old, if not older. There was a generation between them and us.

So we had excellent professors to teach us the mathematics of let us say up to 1900, but we did not know very much about the mathematics of 1920. As I said before, the Germans went about things in a different way, so that the German mathematics school in the years following the war had a brilliance which was altogether exceptional. We only need to think of the mathematicians of the highest order who illustrated this point: C. L. Siegel, E. Noether, E. Artin, W. Krull, H. Hasse, etc., of whom we in France knew nothing.

The prodigious Jacques Herbrand (1908-1931), having studied in Paris under Ernest Vessiot (1865-1952), and having astonished the mathematical world with his findings in metamathematics, came to Göttingen. Among his many interests was ideal theory, and for this reason he was in contact with Emmy Noether in June and July, 1931.

While in Germany, Herbrand lectured to mathematical audiences in Berlin, Halle, and Göttingen. Emmy Noether was especially fond of Herbrand, not only for his scientific abilities, but also for his personal charm and character. To her and many others it was painful news that they would see no more of Herbrand after he left Göttingen in July 1931 for a vacation. Several weeks after the accident, Emmy Noether was moved to write (Dick 1970) "Mir geht der Tod von Herbrand nicht aus dem Sinn." (*The death of Herbrand will not leave me.*)

Herbrand had left some of his work with Emmy Noether which she was able to arrange for publication (Herbrand 1932). Her prefatory note[24] is translated as follows:

Jacques Herbrand, born on the 12th of February, 1908, in Paris, was killed on the 27th of July, 1931, while mountain-climbing in the French Alps. One of the strongest mathematical talents has passed on with him, just during a time of the most intensive work, while he was full of ideas for the future. The last year of his life, which he spent with a Rockefeller scholarship in Germany, brought him into close contact, both scientifically and as a human being, with a number of German mathematicians . . .

Under the initiative of Chevalley and André Weil, friends of Jacques Herbrand honored his memory with a series of works in the French journal *Actualités Scientifiques et Industrielles.*[25] Included among these is the last paper by Emmy Noether [43] to be published during her lifetime.

Aside from her students, Emmy Noether exerted a strong influence on F. K. Schmidt, via Krull. Emil Artin (1898-1962) and Helmut Hasse (1898-1979) stood beside her, Weyl (1935) wrote,

> as two independent minds whose field of production touches on hers closely, though both have a stronger arithmetical texture. With Hasse above all she collaborated very closely during her last years. From different sides, Richard Brauer and she dealt with the profounder structural problems of algebras, she in a more abstract spirit, Brauer, educated in the school of great algebraist I. Schur, more concretely operating with matrices and representations of groups; this, too, led to an extremely fertile cooperation.

After leaving Göttingen, Emmy Noether directed the work of one more doctoral candidate. This was Ruth Stauffer (McKee), who had essentially finished her dissertation *The Construction of a Normal Basis in a Separable Normal Extension Field* (Stauffer 1936) before her adviser's death. Miss Stauffer took her final doctoral examination under Richard Brauer and received her Ph.D. shortly thereafter from Bryn Mawr College.

The last letter which Hasse received from Emmy Noether, dated April 7, 1935, was almost totally mathematical in content, and it dealt primarily with the results which Ruth Stauffer had obtained for her dissertation.

The dissertation was published in 1936, and during the same year, another member of the Bryn Mawr "Noether family," Marie J. Weiss (1936), published an abstract of work which bears the influence of Emmy Noether.

Marie Weiss died before the publication of the second edition of her widely used and influential textbook *Higher Algebra for the Undergraduate* (1949). Writing as Professor of Mathematics at Newcomb College of Tulane University, she acknowledges, in the Preface of her book, the influence especially of van der Waerden's *Moderne Algebra* (1931), A. A. Albert's *Modern Higher Algebra* (1937), and H. Hasse's *Höhere Algebra* (1951).

As van der Waerden's *Moderne Algebra* has served as the most important single carrier of the influence of Emmy Noether, it is worthwhile to compare it with other algebra books. "Among earlier German texts in algebra," Saunders Mac Lane wrote to the author,[26] "the dominant one was *Lehrbuch der Algebra* by Heinrich Weber."

> There was a subsequent useful book by Oscar Perron with the same title, while Robert Fricke wrote a three volume such book which was intended to be the successor to Weber's book. None of these represented the abstract view. However, in 1929, Otto Haupt published a two volume work called *Einfuhrung in die*

Algebra. This was deliberately couched in abstract language but was quite a bit clumsier than van der Waerden. The preface, page 7, explicitly acknowledges the help of Noether. At the suggestion of my teacher, Oystein Ore at Yale University, I studied carefully both volumes of this text by Haupt. I have never known any other algebraist who looked at it. It was, of course, superseded by the much more effective book of van der Waerden.

In some ways, van der Waerden's *Moderne Algebra* (1931) served as a model for the highly influential work of Bourbaki[27] (Dieudonné 1970):

> It is true that there were already excellent monographs at the time and, in fact, the Bourbaki treatise was modelled in the beginning on the excellent algebra treatise of Van der Waerden. I have no wish to detract from his merit, but as you know, he himself says in his preface that really his treatise had several authors, including E. Noether and E. Artin, so that it was a bit of an early Bourbaki ... Since then, algebra has developed considerably, partly because of Van der Waerden's treatise, which is still an excellent introduction. I am often asked for advice on how to start out studying algebra, and to most people I say: First read Van der Waerden, in spite of what has been done since.

Aside from van der Waerden's *Moderne Algebra* (1931) (available in English since 1949), Krull's *Idealtheorie* (1968), Jacobson's *The Theory of Rings* (1943) and *Lectures in Abstract Algebra* (1951), Deuring's *Algebren* (1968), Northcott's *Ideal Theory* (1953), and Zariski and Samuel's *Commutative Algebra* (1958, 1960) are among the foremost and widely accessible expositions of the mathematics of Emmy Noether and her school.

In his letter to the New York Times, Einstein (1935) wrote of Emmy Noether's influence within the totality of human experience:

> The efforts of most human beings are consumed in the struggle for their daily bread, but most of those who are, either through fortune or some special gift, relieved of this struggle are largely absorbed in further improving their worldly lot. Beneath the effort directed toward the accumulation of worldy goods lies all too frequently the illusion that this is the most substantial and desirable end to be achieved; but there is, fortunately, a minority composed of those who recognize early in their lives that the most beautiful and satisfying experiences open to humankind are

not derived from the outside, but are bound up with the development of the individual's own feeling, thinking and acting. The genuine artists, investigators and thinkers have always been persons of this kind. However inconspicuously the life of these individuals runs its course, none the less the fruits of their endeavors are the most valuable contributions which one generation can make to its successors.

NOTES

1. Auguste Dick's excellent monograph (1970) is cited explicitly only a few times. In many cases, however, biographical information in this chapter which appears without a citation was first published in her monograph.
2. Charlotte Angas Scott (1858-1931), having received her Ph.D. in 1885 from the University of London, arrived in the same year at the newly founded Bryn Mawr College, where she inaugurated both the undergraduate and graduate programs in mathematics. "When she retired in 1925 from the department she had directed for forty years, the college's board of directors assessed her influence over those years as being second only to that of President [M. Carey] Thomas herself" (Lehr 1971). For a number of years, Charlotte Angas Scott was an editor of the American Journal of Mathematics, the first mathematical journal to be published in America. In 1904 this journal published an excellent full-page photograph of Max Noether.
3. A discussion with references to Gordan's original publications is given by Kline (1971, p. 929ff).
4. Ernst Fischer is remembered much more for his work in early functional analysis than for contributions to algebra. In 1907 he introduced the concept of mean convergence for Lebesgue square-integrable functions, and this led to the well-known Riesz-Fischer theorem. Fischer remained at Erlangen, except for military service during the war, until 1920. Then he moved to Cologne, where he was a professor until he was prematurely retired in 1938. He was reinstated in 1945 and continued lecturing until the summer before his death.
5. Translation: "*Yesterday I received from Miss Noether a very interesting paper on invariant forms. I am impressed that one can comprehend these matters from so general a viewpoint. It would not have done the Old Guard at Göttingen any harm, had they picked up a thing or two from her. She certainly knows what she is doing.*"
6. Letter to the author, September 13, 1968; reprinted from Kimberling (1972).
7. This is the title of Chapter 12 in van der Waerden's *Moderne Algebra* (1931).

8. Translation: "You can find that already in Dedekind," a dismissal of ideas other may have thought new but which could be found in Dedekind, especially the works reprinted in 1964.
9. Emmy Noether's influence in algebraic topology, via Hopf and Alexandroff, is discussed by Birkhoff (1976, p. 72). See also Behnke and Hirzebruch (1972).
10. Handwritten letters dated 6/1/27 and 7/3/27 from Emmy Noether to W. E. Tisdale supporting Hopf's application are preserved in the International Education Board Collection at the Rockefeller Archive Center. In another file at the Center there is a note that, according to B. L. van der Waerden, when Alexandroff was in Russia he was required to teach for 10 to 13 hours a day.
11. The Brauer-Hasse-Noether theorem is discussed by Feit (1979). See also the note following the *Zentralblatt* reference.
12. Another example is illustrated by the concluding remark in the paper by Hawkins (1972): "If DEDEKIND had not decided to introduce and study group determinants—a subject really outside his main interests in algebraic number theory—or if he had not decided to communicate his ideas on group determinants to FROBENIUS, it is unlikely that FROBENIUS would have created the theory of group character and representations." Emmy Noether contributed [35] to later developments of this theory. See Boerner (1963) and Curtis and Reiner (1962).
13. Ernst Witt.
14. This Institute, financed jointly by the Carnegie and Rockefeller Foundations, through Edward R. Murrow's role as assistant director of the Institute and secretary of the Emergency Committee, was instrumental in bringing almost 300 German scholars to safety. Murrow joined the CBS Radio Network in 1935 and became "the best known and most influential American commentator on World War II." His radio program *Hear It Now* and television program *See It Now* are remembered by millions.

    Extracts in this section are, unless noted otherwise, from the Rockefeller Archive Center and the Bryn Mawr College Archives.
15. Now Grace Shover Quinn, Professor of Mathematics and Statistics (retired), American University.
16. Actually, for only one year.
17. Five of Zorn's works in algebra are listed in the Bibliography of Jacobson (1943).
18. Conversation with the author, June 8, 1980.
19. Copies of the letters by Lefschetz, Wiener, and Birkhoff were found in the Einstein papers at Princeton and in other collections. Included with the Princeton copies is, in addition, a copy of a letter by Einstein. A second copy, preserved in the records of the Emergency Committee (in the New York City Public Library), shows that the letter was dated January 8,

1935 and addressed to Jacob Billikopf. The text and translation are as follows:

> Fräulein Dr. Emmy Noether besitzt unzweifelhaft erhebliches schöpferisches Talent, was jeweilen von nicht sehr vielen Mathematikern einer Generation mit Recht gesagt werden kann. Ihr die Fortsetzung der wissenschaftlichen Arbeit zu ermöglichen, bedeutet nach meiner Ansicht die Erfüllung einer Ehrenpflicht und wirkliche Förderung wissenschaftlicher Forschung.

> Translation: *Miss Dr. Emmy Noether undoubtedly possesses great creative talent, to an extent which cannot be justifiably said about many mathematicians of a generation. To enable her to continue her scientific work means, in my opinion, the fulfilling of a duty of honor and a genuine advancement of scientific research.*

20. Letter, James L. Richards, M.D., to President Park, April 24, 1935.
21. Letter, Dr. David Riesman to President Park, April 18, 1935.
22. Quoted with permission of the MIT Press; copyright 1956 by Norbert Wiener.
23. Information about Chiungtze Tsen was kindly furnished to the author by Professor Hsio-Fu Tuan of Peking University, in a letter dated March 12, 1980.
24. This note, in Emmy Noether's handwriting, is preserved at the University of Evansville.
25. Facing the first paper of the series, by H. Hasse (No. 70, 1934), is an excellent full-page photograph of Herbrand, taken by Emil Artin.
26. In a letter dated November 6, 1979.
27. This is the name of a varying set of writers, consisting initially, it is thought, of C. Chevalley, J. Delsarte, J. Dieudonné, and A. Weil.

ACKNOWLEDGMENTS

Among the many who helped in the gathering and interpreting of materials for this chapter were Nancy Anderson, Head Librarian of the Mathematics Library at the University of Illinois; J. William Hess, Associate Director of the Rockefeller Archive Center; Henry Miner, Professor of German at the University of Evansville; and Gertrude Reed, Head Reference Librarian and Archivist at Bryn Mawr College. For their contributions through correspondence and interviews I also name Mrs. Natascha Artin Brunswick, Dr. Auguste Dick, Dr. Herman Noether, and Professors Garrett Birkhoff, Duane Broline, Nathan Jacobson, Patricia Kenschaft, Marguerite Lehr, Saunders Mac Lane, Otto Neugebauer, Gottfried Noether, Grace Shover Quinn, Earl Taft, and B. L. van der Waerden.

I am deeply grateful.

REFERENCES

Albert, A. A. (1937). *Modern Higher Algebra,* University of Chicago Press, Chicago.

Alexandroff, P. S. (1936). In memory of Emmy Noether, *Uspekhi Mat. Nauk 2,* 254-266. (An English translation appears as Chapter 5 of this book.)

─────── (1969). Die Topologie in und um Holland in den Jahren 1920-1930, *Nieuw Archief voor Wiskunde* (3) *XVII,* 109-127. (Along with his recollection of the reconciliation of Hilbert and Brouwer which he engineered in Emmy Noether's "cozy attic room," the author also mentions that "Edmund Landau was in the habit of asking whether Euler's polyhedron formula was valid for *this* room." At that time, Emmy Noether resided at 57 Friedländerweg. In his paper of 1936, Alexandroff tells the circumstances under which she was forced to find a different residence. Just before she left Göttingen, her apartment was at 51 Stegemühlenweg.)

Amitsur, S. A. (1974). Jacob Levitski 1904-1956, *Israel Journal of Mathematics 19,* 1-3. [The author of this tribute writes, "Levitski actually went to Göttingen with the intention of studying chemistry, but was persuaded by a friend, the late Prof. Benjamin Amirà, to listen to one of Emmy Noether's lectures and from that moment never left mathematics. (His son returned to chemistry.)"]

Baringa, J. (1935). Obituary of Emmy Noether, *Revista Mathematica Hispano-Americana* (2) *10,* 162-163.

Behnke, H., and F. Hirzebruch (1972). In memoriam Heinz Hopf, *Math. Ann. 196,* 1-7.

Berra, A. S. (1935). Obituary of Emmy Noether, *Publicaciones de la Universitad nacional de La Plata 104,* 95-96.

Berzolari, L. (1906). Allgemeine Theorie der höheren ebenen algebraischen Kurven, in *Enzyklopädie der Math. Wiss., III,* 2 Teil, 1 hälfte, Leipzig, 313-455.

Beyerchen, A. D. (1977). *Scientists under Hitler: Politics and the Physics Community in the Third Reich,* Yale University Press, New Haven and London.

Birkhoff, G. (1973). Current trends in algebra, *Amer. Math. Mon. 80,* 760-782; correction *81* (1974) 746. (This award-winning exposition, aside from its emphasis on trends in the early 1970s, gives a brief history of algebra in which van der Warden's *Moderne Algebra,* based largely on lectures of E. Artin and E. Noether, is given special attention.)

─────── (1976). The rise of modern algebra, in *Men and Institutions in American Mathematics,* Graduate Studies, Texas Tech University, No. 13, 41-85. (This is probably the most thorough published treatment of its subject. Pages 41-63 cover *The rise of modern algebra to 1936,* and pages 65-85 cover 1936 to 1950. The "tidal wave" of modern algebra, Birkhoff writes, "was generated in Germany and assumed a coherent shape in Göttingen during the 1920's under the leadership of Emmy Noether." Birkhoff fits the wave into its place in the history of modern mathematics. "In retrospect," Birkhoff writes, "it is easy to recognize the enormous importance of Emmy Noether's work and influence . . . so great that it is easy to overlook the equally fundamental influence of English- and French-speaking mathematicians on the development of 'modern' algebra.")

Boerner, H. (1963). *Representations of Groups*, North-Holland Publishing Company, Amsterdam. [Chapters VII (*Characters and Single-valued Representations of the Rotation Group*) and VIII (*Spin Representations, Infinitesimal Ring, Ordinary Rotation Group*) are credited to Emmy Noether, Hermann Weyl, B. L. van der Waerden, I. Schur, R. Brauer, and others. Regrettably, this crediting is found only in the index: the two chapters do not mention Emmy Noether, and only her paper [34] is included in the five-page bibliography.]

Bourbaki, N. (1960). *Elements d'histoire des mathématiques*, Paris.

Brill, A. (1923). Max Noether, *Jahresber. Deutsch. Math.-Verein* 32, 211-233. (A footnote states that a written account by Emmy Noether was the source of some of the personal information about Max Noether. Included is the following: "His relationship to his mother, who died at the beginning of the 1870's, appears to have been very close; and a brother of his mother, a merchant who was also a sort of private scholar, undoubtedly had mathematical talent, although it was not developed.")

von Brill, A., and M. Noether (1894). Die Entwicklung der Theorie der algebraischen Funktionen in älterer und neuerer Zeit, *Jahresber. Deutsch. Math.-Verein.* 3, 107-566.

Castelnuovo, G., F. Enriques, and F. Severi (1925). Max Noether, *Math. Ann.* 93, 161-181. (The three authors were renowned Italian geometers who wrote, at the request of the editors of the *Annalen*, this commemoration of one "whom we consider one of our most important masters." Footnotes state that Signorina Dr. Emmy Noether supplied the biographical information and the list of 78 publications.)

Chee, P. S. (1975). Emmy Noether—an energetic washerwoman, *Bull. Malaysian Math. Soc.* 6, No. 3, 1-9.

Chevalley, C. (1933). La théorie du symbole de restes normiques, *J. reine angrew. Math.* 169, 140-157.

_____ (1951). Review of *Idealdifferentiation und differente* [44], *Math. Reviews* 12, 388-389. (Heinrich Grell brought about the posthumous publication of Emmy Noether's paper [44]. Chevalley's review, in English, concludes with the remark that "the present paper [whose content was first made public in 1929 at the Prague meeting of the

Deutsche Mathematische-Vereinigung], although published fifteen years after the death of the author, proves once more how much a forerunner she was in all branches of algebra.")

Curtis, C. W., and I. Reiner (1962). *Representation Theory of Finite Groups and Associative Algebras,* Interscience, New York. (The Preface defines representation theory as "the study of concrete realizations of the axiomatic systems of abstract algebra." The first stage in its development was carried out largely by G. Frobenius. [See Hawkins (1972).] "The second state . . . , initiated by E. Noether in 1929, resulted in the absorption of the theory into the study of modules over rings and algebras." [See also Boerner (1963)].)

Dedekind, R. (1930-1932). *Gesammelte mathematische Werke,* herausgegeben von R. Fricke, E. Noether, O. Ore. Vol. I-III. Vieweg, Braunschweig. Also Chelsea, New York, 1969.

——— (1964). *Über die Theorie der ganzen algebraischen Zahlen,* Vieweg, Braunschweig. (The introduction to this book was written by B. L. van der Waerden: ". . . For Emmy Noehter the eleventh supplenent [to the number-theoretic lectures of Dirichlet] was an inexhaustible source of stimulation and methods." It was this supplement which she had in mind when she so frequently said, "Es steht schon bei Dedekind.")

Deuring, M. (1968). *Algebren,* Ergebnisse der Mathematik und ihrer Grenzgebiete, Springer, Berlin. Also Chelsea, 1948.

Dick, A. (1970). *Emmy Noether, 1881-1935,* Birkhäuser Verlag, Basel. English translation by Heidi Blocher, Birkhäuser, Basel, 1981. [Written in German, this monograph is the result of years of research, including travel to Erlangen and Göttingen and discussions and correspondence with many who had close contact with Emmy Noether, including Helmut Hasse, B. L. van der Waerden, Elisabeth Fischer, Max Deuring, Heinrich Grell, Otto Haupt, Wolfgang Krull, Alexander Ostrowski, and Gottfried Noether. The text of 36 pages is accompanied by a 1933 photograph of Emmy Noether, a Chronology, and reprintings of van der Waerden's *Nachruf* (1935) and Weyl's memorial address (1935). The English translation contains several more photographs.]

Dieudonné, J. A. (1970). The work of Nicholas Bourbaki, *Amer. Math. Mon.* 77, 134-145.

——— (1972). The historical development of algebraic geometry, *Amer. Math. Mon.* 79, 827-866. [In this coverage from about 400 B.C. to 1971 A.D., one finds in Section VI (*Development and Chaos 1866-1920*) a treatment of the contributions of M. Noether, A. Brill, and the Italian school. In the next section, one reads, ". . . a thorough examination of the basic concepts from the exclusive viewpoint of algebra, was necessary . . . This groundwork, which at the same time created most of modern commutative algebra, was chiefly due to E. Noether, W. Krull, van der Waerden, and F. K. Schmidt in the period 1920-1940, and to Zariski and A. Weil from 1940 on."]

Dresden, A. (1942). The migration of mathematicians, *Amer. Math. Mon. 49*, 415-429. (This paper gives a year-by-year list of mathematicians who emigrated from Europe between 1933 and 1942. The 10 who came to America in 1933 were Felix Bernstein, Salomon Bochner, Richard Brauer, Albert Einstein, Fritz Herzog, Hans Lewy, Walter Mayer, Emmy Noether, Otto Szász, and Hermann Weyl.)

Dubreil-Jacotin, M-L. (1961). Emmy Noether 1882-1935, in *Les femmes célebrès*, Tome II, Lucien Mazenod, Paris, p. 22.

Einstein, A. (1935). Letter to the editor, *The New York Times*, May 4, 1935. (According to one source, Weyl sent an obituary note to the Times, and the Times reacted, "Who is Weyl—have Einstein write something, as he is the mathematician recognized by the world." There was, at the time, some indignation about the Times' attitude regarding Weyl.)

Encyclopedic Dictionary of Mathematics, Mathematical Society of Japan. English translation: MIT Press, Cambridge, 1977. (Section 281, entitled *Noetherian Rings*, gives basic definitions and concise statements of main theorems, including Cohen's theorem, the Hilbert basis theorem, the Artin-Rees lemma, Krull's intersection theorem, and Krull's altitude theorem. Subsection 281.G is a historical sketch which begins: "J.W.R. Dedekind first introduced the concept of ideals in the theory of integers. The main objects studied in ring theory were subrings of number fields or function fields until M. Sono [*Mem. Coll. Sci. Univ. Kyoto 2* (1917), *3* (1918-1919)] originated an abstract study of Dedekind domains, which was followed by E. Noether [*Math. Ann. 83* (1921), *96* (1926)], who originated the theory of Noetherian rings." Sono and Noether are similarly mentioned together also in Section 10. Section 14.B (*Algebraic Geometry, History*) describes Max Noether's research involving Cremona transformations and the Riemann-Roch theorem, which contributed to the theory of algebraic curves. His influence on the Italian school of algebraic geometry is described briefly. The section also reports that "E. Noether constructed an abstract theory of polynomial ideals from a formal theory by E. Lasker and F. S. Macaulay. Under her influence there appeared the arithmetic algebraic geometry (of curves) over an abstract field as developed by F. K. Schmidt and others." Section 30.D contains a statement of the Hasse-Brauer-Noether theorem [39], and Section 30.E juxtaposes it with Tsen's theorem.)

Evans, R. J. (1976). *The Feminist Movement in Germany 1894-1933*, Sage Publications, London and Beverly Hills.

Feit, W. (1979). Richard D. Brauer, *Bull. Amer. Math. Soc.* (New Series) *1*, 1-20. (This tribute includes discussions of the papers which Richard Brauer wrote with Emmy Noether [33] and with her and Helmut Hasse

Emmy Noether and Her Influence

[39]. "He was a mathematician in the tradition of Hermann Weyl. Although he collaborated with Emmy Noether, he was perhaps less influenced by her abstract point of view than any other German algebraist of his generation.")

Fisher, C. S. (1967). Death of a Mathematical Theory: a Study in the Sociology of Knowledge, *Arch. Hist. Exact Sci. 3*, 136-159. (Writing from the Department of Sociology and Anthropology at Princeton University, the author traces the history of invariant theory. Work by Emmy Noether is "seen not so much as a change in Invariant Theory proper, but more as a carrying of a result which was previously thought to be set in Invariant Theory into another part of mathematics.")

Fogarty, J. (1969). *Invariant Theory*, W. A. Benjamin, Inc., New York. (This book presents the major contributions to algebraic invariant theory of Hilbert, Emmy Noether, Weitzenbock, Fischer, Weyl, and others. It also offers insights of historical interest. For example: "It is of more than passing interest to note that the origins of the theory of noetherian rings can be found in Hilbert's work on invariant theory.")

Jahrbuch über die Fortschritte der Mathematik [From 1868 to 1942, this publication served the same purposes as *Mathematical Reviews* since 1940 and *Zentralblatt für Mathematik und Ihre Grenzgebiete* since 1931, that is, to summarize recent research and other information of interest to mathematical scientists. Most of the publications of Emmy Noether (as numbered at the end of this book) are summarized in *Fortschritte*, including the following:

Publication	*Fortschritte* vol. and yr.	Pages	Reviewer
1 and 2	39 1908	158-160	Meyer (Königsberg)
3 and 4	41 1910	149-152	Meyer
6	46 1916-18	1442-1443	E. Noether
7	45 1914-15	198-199	Meyer
8	45 1914-15	197-198	Meyer
9	45 1914-15	163-164	F. Jacobsthal (Berlin)
10	46 1916-18	170	E. Noether
11	46 1916-18	135	E. Noether
12	46 1916-18	675	E. Noether
13	46 1916-18	770	E. Noether
14	47 1919-20	349-350	E. Noether
16	47 1919-20	89	E. Noether
17	47 1919-20	97-98	E. Noether
19	48 1921-22	121	E. Noether
20	48 1921-22	81	W. Schmeidler (Breslau)
22	48 1921-22	94-95	W. Schmeidler
23	49 1923	68	E. Noether
24	49 1923	73-77	W. Schmeidler
25	50 1924	71	W. Schmeidler

Publication	Fortschritte vol. and yr.	Pages	Reviewer
30	52 1926	106	B. H. Neumann
31	52 1926	130-132	E. Pannwitz (Berlin)
32	52 1926	125	Sperling
33	53 1927	116	A. Loewy (Freiburg)
35	55 1929	677-678	Sperling
36	55 1929	666-667	W. Specht (Breslau)
38	58 1932	172-173	A. Lotze (Stuttgart)
39	58 1932	142-143	A. Lotze
40	58 1932	142	E. Trost (Zurich)
41	59 1933	152-153	H. Hasse (Marburg)
42	59 1933	941-942	H. Hasse
43	60 1934	101-102	H. Reichardt (Berlin)

Emmy Noether wrote summaries for *Fortschritte* which appeared during the years 1916-1928. Among those whose work she reviewed were W. Schmeidler (v. 46), K. Hensel (v. 47), A. Hurwitz (v. 47), I. Schur (v. 48), G. Szegö (v. 48), A. Loewy (v. 48), E. Fischer (v. 49), C. C. MacDuffee (v. 49), F. S. Macaulay (v. 53), H. Grell (v. 53), R. Hölzer (v. 53), E. T. Bell (v. 53), and B. L. van der Waerden (v. 54). The lengthy review of Ref. 31 was written by Dr. Erika Pannwitz (1904-1975) during a time when there were still very few women mathematicians in Germany. After study in Berlin, Freiburg, and Göttingen, she received her Ph.D. under Heinz Hopf in Berlin in 1931. Dr. Pannwitz also reviewed the Cantor-Dedekind volume (Noether and Cavaillès 1937) in *Fortschritte 63* (1937) 18-19. *Fortschritte* also contains summaries of many publications of Max Noether and Fritz Noether.]

Grinstein, L. S., and P. J. Campbell (1978). Anna Johnson Pell Wheeler, *Association for Women in Mathematics Newsletter,* Sept. 1978 and Nov. 1978.

Hasse, H. (1932). Theory of cyclic algebras over an algebraic number field, *Trans. Am. Math. Soc. 34,* 171-214.

———— (1933). Die Struktur der R. Brauerschen Algebrenklassengruppe über einem algebraischen Zahlkörper, *Math. Ann. 107,* 731-760.

———— (1951). *Höhere Algebra,* third ed., Walter de Gruyter, Berlin.

Hawkins, T. (1972). Hypercomplex numbers, Lie groups, and the creation of group representation theory, *Archive Hist. Exact Sci. 8,* 243-287.

Herbrand, J. (1932). Théorie arithmétique des corps de nombres de degré infini, *Math. Ann. 106,* 473-501. (See also p. 502.)

Houtappel, R. M. F., H. Van Dam, and E. P. Wigner (1965). The Conceptual Bases and Use of the Geometric Invariance Principles, *Rev. Mod. Physics 37,* 595-632. (This paper gives the historical and physical context of the famous "Noether's theorem" which Emmy Noether proved in her Habilitationsschrift [13].)

Humphreys, J. E. (1978). Hilbert's Fourteenth Problem, *Amer. Math. Mon. 85*, 341-353. [This paper traces the solution of certain important cases of Hilbert's Fourteenth Problem (from the famous 23 problems which Hilbert presented to the Second International Congress of Mathematicians at Paris in 1900) back to Emmy Noether's papers [7] and [30]. (The 23 problems, with comments, are published in English in Section 198 of the Encyclopedic Dictionary of Mathematics.)]

Ibragimov, N. Kh. (1972). *Lie Groups in Problems of Mathematical Physics*, Novosibirsk (in Russian). [The author, of the Hydrodynamics Institute, Siberian Branch, Academy of Sciences of the U.S.S.R., here gives a detailed discussion of the questions related to Noether's theorem (from Ref. 13) and its extensions.]

Men of Modern Mathematics (1966). Wall chart produced for International Business Machines (IBM) by the office of Charles Eames; biographies and mathematical notes by Ray Redheffer. [Among the men pictured and described are Dedekind, Hilbert, Weyl, Wiener, and Artin. (Klein is missing.) The only woman included is Emmy Noether. This large and impressive chart is available from the Museum of Science and Industry, Chicago.]

Jacobson, N. (1943). *The Theory of Rings*, American Math. Soc. Mathematical Surveys, No. 2, Providence. Also 1968.

———— (1951). *Lectures in Abstract Algebra*, Vols. I, II, III, Van Nostrand, Princeton. [In Volume I, the fifth chapter, *Groups with Operators*, is introduced with the statement that "this concept ... was first considered by Krull and Emmy Noether." Similarly, in the introduction to the next chapter, *Modules and Ideals*, one finds: "Modules are of fundamental importance in the study of homomorphisms of abstract rings into rings of endomorphisms of commutative groups (so-called representation theory). This was first recognized by Emmy Noether." In Volume III, the first chapter, *Finite Dimensional Extension Fields*, contains a section entitled *Galois cohomology*. Here *Emmy Noether's equations* ($\mu_{st} = \mu_s^t \mu_t$) are discussed with applications to Hilbert's Satz 90.]

Kaplansky, I. (1970). *Commutative Rings*, Allyn and Bacon, Boston.

Kimberling, C. (1972). Emmy Noether, *Amer. Math. Mon. 79*, 136-149.

Klein, F. (1893). Vergleichende Betrachtungen über neuer geometrische Forschungen, *Math. Ann. 43*, 63-100. [An English translation of this work, which first introduced Klein's famous Erlangen program in an 1872 speech, is published in the *New York Math. Soc. Bull. 2* (1893), 215-249.]

Kline, M. (1972). *Mathematical Thought from Ancient to Modern Times*, Oxford Univ. Press, New York. (Chapter 39, *Algebraic Geometry*, discusses the contributions of Clebsch, Gordan, Max Noether, and others. Documentation of original sources is conveniently given in footnotes. Chapter 49, *The Emergence of Abstract Algebra*, presents an equally useful discussion up to a partial summary of Emmy Noether's contributions.)

Kořínek, V. (1935). Obituary of Emmy Noether, *Časopis pro pěstování matematiky a fysiky 65*, D1-D6.

Krull, W. (1968). *Idealtheorie,* Ergebnisse der Mathematik und ihrer Grenzgebiete, 2nd ed., Springer, Berlin. (First published in 1935, this is the standard work on ideal theory before 1935. Accordingly, it is one of the most important representations of the work of Emmy Noether and her school of algebra.)

Lehr, M. (1971). Scott, Charlotte Angas, in *Notable American Women, 1607-1950,* Belnap Press of Harvard University Press, pp. 249-250. [Miss Lehr was Charlotte A. Scott's last Bryn Mawr Ph.D. (1925) and was on the faculty of Bryn Mawr when Emmy Noether was there. In a private communication to the author in 1968, she wrote, "I coveted (being interested in Algebraic Geometry) a most beautifully illuminated scroll which hung, framed, on her wall. It has been designed for and presented to her father, ... and it showed the "Tree of Algebraic Geometry" from its early roots, with many branchings, all labelled with the quite international roster of illustrious names—one of the current and flowering ones being Noether himself." In the same letter Miss Lehr wrote, "My personal feeling is that Miss Noether, like another algebraist, the brilliantly gifted Emil Artin, influenced mathematics not so much by any great volume of published results, as by a kind of catalytic effect on young vigorous minds. Where they walked and talked, work came about—individual and fresh."]

Logan, J. D. (1977). *Invariant Variational Principles,* Academic Press, New York. (The Preface states, "Since the early part of this century when Emmy Noether wrote her monumental paper on the subject, invariant variational principles have attained extensive application to a wide range of problems in physics and engineering; indeed, the Noether theorem has become one of the basic building blocks of modern field theories. The goal of this monograph is to transmit the flavor of some of these problems and applications and to present some concrete examples which it is hoped will lead the reader to appreciate the connection between invariance transformations and conservation laws for physical systems." The monograph includes ample historical notes, sections of exercises, and a four-page list of references, most of which treat one or both of the Noether theorems first published in Ref. 13.)

Mac Lane, S., Origin, Rise, and Decline of Abstract Algebra, in *American Mathematical Heritage: Algebra and Applied Mathematics,* Graduate Studies, Texas Tech University, in press.

Mathematische Annalen [First published in 1869 under the editorship of A. Clebsch and C. Neumann, this journal published many mathematical works of Max Noether, Emmy Noether, and Paul Gordan. During the years that invariant theory was "Trumpf," *Math. Ann.* was the foremost journal for invariant-theoretic research. It continues to be a leading mathematical journal. Hermann Weyl wrote, "Emmy Noether was a zealous collaborator in the editing of the *Mathematische Annalen*. That her work was never explicitly recognized may have caused her some pain." In his 1965 book, Heinrich Tietze points out in a footnote (p. 180) that he received a copy of Weyl's memorial address on Emmy Noether, "and its warmth touched me deeply. But van der Waerden's memorial article appeared in Germany; its author was at that time in Leipzig. Those who knew the situation in Germany at the time, or can imagine it, will be able to appreciate fully the courage of the author of the article and of the editors of the *Annalen*."]

May, K. O. (1972). Review of F. A. Medvedev, "Richard Dedekind," *Math. Reviews 43*, 336-337.

Nagao, H. (1978). Kenjiro Shoda, 1902-1977, *Osaka J. Math. 15*, i-iv.

Noether, E., and J. Cavaillès, editors (1937). *Briefwechsel Cantor-Dedekind*, Actualités Scientifiques et Industrielles, 518, Hermann, Paris.

Noether, E. P. (1976). Emmy Noether (1882-1935), speech sponsored by the Association for Women in Mathematics, Joint Mathematics Meetings, Toronto, 1976. Also [with Smith (1976)] in *AWM Newsletter 6*, No. 7. [The author is a Professor of History at the University of Connecticut. Her husband, Gottfried Noether, the younger son of Fritz Noether, is Professor and Head of the Department of Statistics at the University of Connecticut. See also Smith (1976).]

Noether, M. (1869). Zur Theorie des eindeutigen Entsprechens Algebraischer Gebilde von beliebig vielen Dimensionen, *Math. Ann. 2*, 293-316.

_____ (1872). Über einen Satz aus der Theorie der algebraischen Funktionen, *Math. Ann. 6*, 351-359.

_____ (1909). Übermittlung von Nachschriften Riemannscher Vorlesungen, *Göttinger Nachr. Geschäftl. Mitt.*, 33-25. (Reprinted by Dover, 1953.)

_____ (1914). Paul Gordan, *Math. Ann. 75*, 1-41.

Northcott, D. C. (1953). *Ideal Theory*, Cambridge Tracts in Mathematics and Mathematical Physics, No. 42, Cambridge University Press, Cambridge.

Osofsky, B. L. (1974). Global Deformation of Polarized Varieties, *Bull. Amer. Math. Soc. 80*, 1-26. (Section 3 of this paper traces the development of homological dimension theory starting with Emmy Noether [19] in 1921.)

Pinl, M., and L. Furtmüller (1973). Mathematicians under Hitler, *Leo Baeck Institute Yearbook*, London, *XVIII*.

Polak, L. S. (1959). *Variatsionnye Printsipi*, Mekhanika Moskva, Gosudarstvennoe Izdatel'stvo, Fiziko-Matematicheskoi Literatury. (Pages 611-630 are a translation into Russian of the 1918 paper, Ref. 13, in which Noether's theorem was first stated and proved.)

Polubarinova-Kochina, P. (1957). *Sophia Vasilyevna Kovalevskaya, Her Life and Work,* Foreign Languages Publishing House, Moscow.

Polyá, G. (1973). A story with a moral, *Math. Gazette* 57, 86-87. (The author recalls a discussion he had with Emmy Noether after a lecture he gave in Göttingen in the early 1930s.)

Reid, C. (1970). *Hilbert,* Springer-Verlag, New York. (In 1969 Richard Courant wrote for the Foreword of *Hilbert*: "I trust that the book will fascinate not only mathematicians but everybody who is interested in the mystery of the origin of great scientists in our society." As in the author's later book, *Courant,* there are a number of paragraphs about Emmy Noether. There are photographs of Emmy Noether, Paul Gordan, Hilbert, Weyl, Minkowski, Klein, and others.)

———— (1976). *Courant in Göttingen and New York,* Springer-Verlag, New York. (Within these 309 pages the scientific and social atmosphere of Göttingen comes alive, especially to readers who have studied the mathematics of Hilbert, Klein, Minkowski, Landau, Artin, Courant, Hasse, Noether, Neugebauer, Runge, Weyl, and others. Emmy Noether frequently appears, and hers is one of the 36 handsome photographs in the Album at the end of the book.)

Ringer, F. K. (1969). *The Decline of the German Mandarins: The German Academic Community, 1890-1933,* Harvard University Press, Cambridge, Massachusetts.

Schilling, O. (1950). *The Theory of Valuations,* Amer. Math. Society Mathematical Surveys, Providence.

Science Citation Index, Institute for Scientific Information, Philadelphia 1965-. [This massive index, often used with the help of a computer, enables one to identify literature which cites any given scientific publication. For the years 1965-1979 the number of mathematics journals indexed varies between 110 and 170, and approximate numbers of citations in these and physics journals of publications of Emmy Noether (as numbered at the end of this book) are as follows:

Publication	1965-69	1970-74	1975-79 (August)
7	1	3	7
11	2	3	2
13	46	64	50
24	6	2	1
38	2	2	5
44	7	1	0
others	11	9	5

Published in conjunction with *Science Citation Index* is the *Subject Permuterm Index,* produced by a computer which permutes significant words within titles and subtitles of scientific journal articles. For a title of n significant words, the computer generates $n(n-1)$ pairs, each consisting of a *primary term* and a *co-term*. For example, the 1978 Permuterm Index lists, under the primary term *Noether,* a total of 29 distinct co-terms; and under the primary term *Noetherian,* a total of 66 distinct co-terms.]

Scott, C. A. (1899). A proof of Noether's fundamental theorem, *Math. Ann. 52,* 593-597.

Segal, S. L. (1980). Helmut Hasse in 1934, *Historia Mathematica 7,* 46-56.

Shoda, K. (1932). *Abstract Algebra,* Iwanami, Tokyo. (Through this book, Emmy Noether's influence was brought to Japan. "When I was a student," wrote Hirosi Nagao, "I heard from [Shoda, a 1927 'Noether boy'] that it was the first book in the world containing the theory of the crossed product by E. Noether.")

Sloane, N. J. A. (1977). Error-correcting codes and invariant theory: new applications of a nineteenth-century technique, *Amer. Math. Monthly 84,* 82-107. (This article offers a very readable introduction to algebraic invariant theory and points out that, although this theory has been unfashionable for some time, it provides here a "technique ... potentially of much wider application, ... very powerful, ... and fun to use." The list of 74 references includes Emmy Noether's paper Ref. 7.)

Smith, M. (1976). Emmy Noether's Contribution to Mathematics, speech sponsored by the Association for Women in Mathematics, Joint Mathematics Meetings, Toronto, 1976. [See also Noether (1976).]

Stauffer, R. (1936). The construction of a normal basis in a separable normal extension field, *Amer. J. Math. 58,* 585-597.

Tavel, M. A. (1971). Noether's Theorem, *Transport Theory and Statistical Physics 1(3),* 183-207. (Under the heading *Milestones in Mathematical Physics,* the author has published a translation of Emmy Noether's paper *Invariante Variationsprobleme* [13]. This was her Habilitationsschrift, published in 1918. Preceding the actual translation are two pages of introduction and one page of related bibliography. The introduction outlines a wide variety of applications of Noether's theorem.)

─────── (1971). Application of Noether's Theorem to the transport equation, *Transport Theory and Statistical Physics 1(4),* 271-285.

Thomas, M. C. (1979). *The Making of a Feminist, Early Journals and Letters of M. Carey Thomas* (M. H. Dobkin, ed.), The Kent State University Press. [The year Emmy Noether was born, Miss Thomas wrote, in one of the many letters published in this book, "I had just prepared myself for an examination at Göttingen and expected to take it in August. I relied upon the letters of a Frau prof. in Göttingen whose husband had brought my case up ... *formally* there *was* no objection, but each of the forty-two profs had to be asked his opinion upon the woman question and if two or three disagreed my fate would be decided in the negative." "The M. Carey Thomas Library building," writes Ms. Dobkin in the Introduction, "exemplifies better than any [other] on the Bryn Mawr campus her urge for perfection, her taste for solid conventional elegance ... The cloisters, too, are laid out exactly as Carey Thomas decreed." (Within the cloisters are buried the ashes of both Carey Thomas and Emmy Noether.)]

Tietze, H. (1965). *Famous Problems of Mathematics* (B. K. Hofstadter and H. Komm, eds.) Graylock Press, Baltimore.

Tsen, C-C. (1936). Zur Stufentheorie der quasi-algebraisch-Abgeschlossenheit kommutativer körper, *J. of the Chinese Math Soc. 1*, 81-92.

Veysey, L. R. (1971). Thomas, Martha Carey, in *Notable American Women*, Belnap Press of Harvard University Press, pp. 446-450.

van der Waerden, B. L. (1931). *Moderne Algebra*, 2 volumes, Springer, Berlin.

──────── (1935). Nachruf auf Emmy Noether, *Math. Ann. 111*, 469-476. (An English translation appears as Chapter 4 of this book.)

──────── (1948). The Foundation of Algebraic Geometry—A Very Incomplete Historical Survey, in *Studies and Essays* (K. O. Friedrichs, O. E. Neugebauer, and F. F. Stoker, eds.), Interscience Publishers, Inc., New York.

──────── (1975). On the sources of my book, *Moderne Algebra*, *Historia Mathematica 2*, 31-40.

Weaver, W. (1970). *Scene of Change, A Lifetime in American Science*, Charles Scribner's Sons, New York. [In this autobiography, Weaver tells that he studied under Max Mason (1877-1961) at the Univesity of Wisconsin, where Weaver later became Chairman of the Mathematics Department. Mason had received his Ph.D. under Hilbert at Göttingen in 1903. Having served as President of the University of Chicago, 1925-1928, Mason became President of the Rockefeller Foundation, a position he held during the years the Foundation provided part of Emmy Noether's salary. "Mason and Weaver" became a common phrase in American applied mathematics, notably as a result of their book *The Electromagnetic Field*. Warren Weaver Hall at New York University is the home of The Courant Institute for Mathematical Sciences. There, a sizable portion of what was once Emmy Noether's personal library, including volumes which once belonged to Max Noether, is preserved.]

Weiss, M. (1936). Fundamental systems of units in normal fields, *Bull. Amer. Math. Soc. 42*, 36.

──────── (1949). *Higher Algebra for the Undergraduate*, Wiley, New York.

Weyl, H. (1935). Emmy Noether, *Scripta Mathematica 3*, 201-220; reprinted in *Gesammelten Abhandlungen*, III, Springer, 1968; and in Dick's book (1970). (This is the published version of Weyl's memorial address at Bryn Mawr College, a tribute to Emmy Noether which has served as a framework for the writing of this chapter.)

Wiener, N. (1956). *I Am a Mathematician*, M.I.T. Press, pp. 162-163.

Wussing, H., and W. Arnold (1978). *Biographien Bedeutender Mathematiker,* Aulis Verlag Deubern & Co., KG, Köln. (First printed in East Germany, 1975.) [The final biographical sketch in this book (pp. 504-513) is about Emmy Noether.]

Zariski, O., and P. Samuel (1958, 1960). *Commutative Algebra,* 2 vols., Van Nostrand Reinhold, New York.

Zentralblatt für Mathematik und Ihre Grenzgebiete. (1931). (Reviews of the publications of Emmy Noether appear in *Zentralblatt* much as in *Fortschritte der Mathematik.* In vol. 3 (1932), Max Deuring reviews Ref. 38 on pp. 146-147, and C. C. MacDuffee reviews Ref. 39 on p. 244. In vol. 7 (1933), Deuring reviews Ref. 40 on pp. 3-4, van der Waerden reviews Ref. 41 on p. 197, and Deuring reviews Ref. 42 on p. 195. In vol. 39 (1951), W. Krull reviews Ref. 44 on pp. 26-27. [See Chevalley's paper (1951) for another review of Ref. 44.])

II

NOETHER AND HER COLLEAGUES

2

Mathematics at the University of Göttingen 1931–1933

Saunders Mac Lane

Emmy Noether's career as an influential member of the mathematical institute at Göttingen came to an end in 1933. I was a student there in the period 1931-1933; on the basis of my recollections, I will endeavor in this article to describe the mathematical activities at Göttingen during this period so as to provide some background for Emmy Noether's accomplishments there.

Göttingen was at that time one of the great world centers of mathematics. There were only a few other centers at a comparable level: Paris, Berlin, and perhaps Moscow. Centers in the United States, such as Princeton, were not then quite up to that standard. In the late 1920s the Rockefeller Foundation had made grants to construct two buildings for mathematics, one at Göttingen and one at Paris: The Institute Henri Poincaré in Paris and the Mathematical Institute at Göttingen (an Institute still used, located right next to the equally famous Institute of Physics in Göttingen).

Since the late 1800s many American mathematicians had studied for the Ph.D. at Göttingen, especially under the guidance of David Hilbert. H. B. Curry (Ph.D. 1930) was, I believe, the last such Hilbert student from the United States.

The mathematical staff at the Institute in Göttingen was small compared to present-day centers of mathematics. Nevertheless, that institute did have a dominant position in mathematics and this article will try to explain the reasons for this dominance.

There were only four full professors (Ordentliche Professoren); the rest of the staff, as best I recollect it, is as listed in Appendix A.

There were perhaps six dominant figures in the mathematical life of Göttingen: Hilbert, Courant, Herglotz, Landau, Weyl, and Noether.

David Hilbert, renowned for his work in invariant theory, foundations of geometry, number theory, integral equations, and logic, had been for many years the leading German mathematician. There had long been a rivalry in German mathematics between Berlin and Göttingen. This may account for the rumor which I recently heard in Berlin. This rumor said that in 1889, there had been a proposal to call Hilbert to a professorship at Berlin; one of the Berlin mathematicians examined Hilbert's work, observed that his contribution to algebraic number theory was more in organization than in original discovery and concluded that Hilbert was not up to the Berlin standard! In Göttingen, Hilbert clearly set the standard.

Emmy Noether had first come to Göttingen to work with Hilbert. In 1931, Hilbert had recently retired; he was no longer in very good health. Once a week he lectured on *Philosophy from the Point of View of the Natural Sciences*. From the lectures I recollect only his enthusiastic remark that the discovery of the North American continent by Columbus represented a major turning point in Western civilization. Hilbert's own work in functional analysis and Hilbert spaces was finished, but the subject was still well represented at Göttingen by younger people, for example, by Franz Rellich (later himself a professor at Göttingen, 1945-1949).

At that time, Volume I of Hilbert's collected papers was being edited, with the assistance of Wilhelm Magnus, Helmut Ulm, and Olga Taussky. She had just come from Vienna where she had studied class field theory with Furtwängler; the knowledge was important because this volume included Hilbert's impressive pioneering work on relative quadratic extensions, which provided the starting point for class field theory.

In 1931 the center of Hilbert's interest was in the foundations of mathematics. Stimulated by the disagreements with Brouwer and by what seemed then to be a "crisis" in the foundations of mathematics, he had formulated a program to prove all of mathematics consistent, using finitary methods. His assistants in this work included Paul Bernays [subsequently professor of the Eidgenossische Technische Hochschule (the ETH) in Zurich] and Arnold Schmidt (later a professor at Marburg). Hilbert had several students in logic, notably Gerhard Gentzen and Kurt Schütte.

Gödel had just proved his incompleteness theorem asserting that in suitable formal systems, as for example in *Principia Mathematica*, there can be no consistency proof which could be formulated (via Gödel numbering) as a theorem within the system. This result was a major challenge to the Hilbert program. It was then felt by Bernays and others that Hilbert's notion of "finite methods" could be stretched to avoid the consequences of Gödel's theorem. Gentzen's subsequent work using transfinite induction to establish consistency of a portion of analysis also goes in this direction. At the time, Gödel was annoyed that Hilbert could publish a paper (on "tertium non datum") without any reference to Gödel's result.

Foundations of mathematics generally attracted very lively attention. At one time, Professor Richard von Mises visited from the University of Berlin to give a lecture on the Foundations of Probability—a lecture in which he expressed his view of probability in terms of a "Kollectiv." Briefly speaking, he considered sequences of zeros and ones which were random in the sense that the limiting ratio of zeros to ones remained the same for a subsequence selected by any rule. The complication of this notion, then and now, lay in the words "by any rule." At the end of von Mises' lecture there was an exceedingly lively discussion in which Hilbert, Bernays, Paul Hertz, and others expressed their sharp disagreements with the approach taken by von Mises. In particular, Bernays was heard to remark, "Aber das ist doch völliger Unsinn."

Richard Courant was Professor at Göttingen as a successor to Felix Klein. Klein had been very influential in German mathematics, late in his career especially in exposition, organizations, and in locating positions for young mathematicians. Courant was likewise influential and held the position of Director of the Mathematical Institute. He frequently taught the course on calculus (*Differential- und Integralrechnung*) which had a large enrollment (professors received supplementary income in proportion to the number of their students). During the year 1932-1933, E. J. McShane and his wife came to Göttingen; he was there as assistant to Courant with the task of preparing an English translation, subsequently published, of Courant's text on calculus.

Courant had clear views about exposition (for example, he never used equation numbers as references, either in his calculus or elsewhere). He was a great enthusiast for mathematics, as I know from a period when I stayed in his house in order to help teach him English. It is my recollection that his views at the time about mathematics were considerably broader than those anti-abstract opinions he espoused during his subsequent influential work at New York University, where his emphasis was very strongly on applied mathematics.

A number of junior people were present as assistants or associates of Courant. For example, Wilhelm Cauer (in whose apartment I also lived) worked on the mathematics of electrical networks. Kurt Friedrichs, another associate of Courant, had recently left Göttingen to become a Professor at the Technische Hochschule in Braunschweig.

Herglotz was a much quieter professor. He had come to Göttingen from Leipzig (where Emil Artin had prepared his Ph.D. thesis as a student of Herglotz). Herglotz had an enormous fund of knowledge about nearly every branch of classical mathematics and regularly gave beautifully organized and polished lectures on a great variety of subjects. They included, in my time, the following (for those starred I have mimeographed lecture notes):

Interpolation and extremal problems of analysis
Lie groups
*Geometrical optics (summer 1932)
*Analytical mechanics (summer 1930, winter 1932, notes by Fritz John)
*Contact transformations (summer 1931)
Elliptic functions (notes by Fritz John)

Herglotz often lectured from 7 to 9 in the evening, holding forth in morning coat and striped trousers and sporting a white piqué vest. The main theorems of each lecture were carefully written on the center blackboard, while the incidental calculations and proofs were relegated to the side boards. Each important point was emphasized with a decisive and elegant gesture. Some of the celebrated lecture style of Artin must have been inherited from Herglotz. Most of the older students and assistants were regular attendants at Herglotz's lectures and several of the students were especially devoted to him—Schwerdtfeger and Scherk, for example. Nevertheless, Herglotz did not direct many Ph.D. theses, though Scherk did write his doctoral thesis under Herglotz.

Herglotz's delight in lecturing was well known. On my final Ph.D. oral examination I had as minor subject geometrical function theory, with examiner Herglotz. Other students told me how to behave and I profited by this advice. At the beginning of the examination, Herglotz asked me if I knew the "Klein Erlanger program" and, if so, what group was necessary to make geometrical function theory fit that program. I mumbled something about the conformal group. This was enough to get Herglotz started lecturing on this interpretation; his elegant presentation filled up the whole time of the examination.

When I first visited Göttingen after the war, in 1948, Herglotz was still there, in poor health but still interested in mathematics and happy to remember the good times when Göttingen was strong.

Edmund Landau had come from Berlin in 1909 to be a professor at Göttingen. There, in a vigorous and decisive style, he presented analytic number theory. His lectures on the zeta function, on Dirichlet series, or on analytic number theory were given in the biggest lecture room where the enormous blackboards moved on rollers. These lectures came in the meticulous "theorem-proof" style also displayed in his books. No discussion of reasons, motivations, or insight marred the rapid fire of theorem and proof in his lectures. They were organized so explicitly that I was easily able to take complete notes, writing not only what Landau said but putting on the side my own interpretation of the motivation. One of his assistants sat at the front of the room and had the task of washing the used blackboards with a wet sponge so that the professor could continue without interruption.

Landau had many assistants and students, including Werner Weber, Hans Heilbronn, and Werner Fenchel. Landau's insistence on a rigorous theorem-proof style came out very forcefully. For example, in one of his papers (1932) he called two theorems "equivalent" because each of them is an easy consequence of other. However, he immediately worried that someone might ask him for a precise definition of "easy." Hence he adds to the word "equivalent" the following footnote (translated): "I call two statements equivalent when either both are correct or both are false!"

There were many other Landau stories: One of them was to the effect that when Hardy was finally scheduled to first visit Göttingen, Landau went to the train to meet him. As soon as Hardy descended from the carriage, Landau started to quiz him about estimations for the Waring problem on the major and minor arcs. To which this Hardy replied that he had given up his interest in number theory! It subsequently appeared that "Hardy" was not Hardy but some Göttingen student disguised as Hardy!

Landau was well-to-do—his wife was related to Ehrlich, the chemist who had invented Salversan, the best drug then for venereal disease. Landau lived in a fine house on one of the best streets in Göttingen; there he gave splendid lively and long parties full of all sorts of competitive and mathematical games.

Hermann Weyl was the premier professor of mathematics at Göttingen. He had recently come there from the ETH at Zürich as successor to Hilbert. His interests in mathematics were wide and incisive. I recall his lectures on differential geometry, on Lie group theory, on the elementary ideas of algebraic topology and on the philosophy of mathematics (*Die Erkenntnistheoretische Methode in Mathematik und Physik*).

In connection with his work on continuous groups he expressed the vigorous opinion that the subject of Lie algebras would soon develop actively —a prognosis that turned out to be correct. He conducted a seminar which I attended and which, like most German seminars, required the students participating to lecture on current papers. From that seminar I learned more about elementary divisor theory and also learned how and why vector spaces over a field could be described by axioms. (From Chicago I had gotten the impression that vectors were n-tuples of numbers and vector spaces were just suitably closed sets of n-tuples.) Weyl's presence and style were quite austere; he lived in an impressive apartment in a hilly part of town. There he and his charming wife gave elegant parties. At the start of each semester, students and assistants could call at the homes of professors. On such occasions, the professors were systematically not at home, but they responded later with invitations to parties, such as those at Weyl's, Landau's, or Courant's.

Weyl's chief assistant was Heinrich Heesch. He was much interested in geometric problems of tiling the plane with irregular shapes. It is the same Heesche who much later thought of the possibility of using computers to tackle the four color problem. It was essentially his design that was subsequently employed by Appel and Haken in their recent solution of this problem.

When I first came to Göttingen I spoke to Professor Weyl and expressed my interest in logic and in algebra. He immediately remarked that in algebra Göttingen was excellently represented by Professor Noether; he recommended that I attend her courses and seminars. Noether had originally come to Göttingen in 1919 on the invitation of Hilbert. By the time of my arrival she was an Ausserordentlicher Professor. However, it was clear that in the view of Weyl, Hilbert, and the others, she was right on the level of any of the full professors (Ordentliche Professoren). Her work was much admired and her influence was widespread.

That first fall I did take her course, which dealt with representations of noncommutative rings, a topic on which she was then busily writing a paper. I believe the paper is Ref. 40.* Her lectures were jerky and very enthusiastic, but also rather obscure. Since she was in the process of working out the ideas of the paper, she did some of the working out right there in class. I am afraid that at the time I was sufficiently put off by this obscurity that I did not attend her subsequent courses. Her courses were sparsely attended but by enthusiastic students, plus some faculty members. I remember in particular that Ernst Witt was then one of her most devoted students, and that Paul Bernays regularly came to her courses. In the intermissions, I walked up and down the hall with Bernays, listening to his views on logic and foundations.

---

*Numbered references are to Noether's papers listed at the end of this book.

One story about Emmy Noether was widely related by Courant and others: The mathematicians at Göttingen at one point brought in a recommendation that Dr. Noether be made a full professor, not just an Ausserordentlicher Professor. This could not be done without the consent of a majority of the whole faculty at Göttingen; this faculty contined a number of people who thought that women should not be professors. During the ensuing arguments, Hilbert got up and said in typical Hilbert East German accent, "Aber meine Herren, die Universität ist doch keine Badeanstalt" (*Gentlemen, the university is not a seaside bathing establishment*). At that time such establishments separated women from men.

Noether's enthusiasm for lecturing was not much impeded by days when the Institute was shut. I recall one day when the Institute was not scheduled to be open because of a state holiday. Noether announced that the class would go on just the same, but would take the form of an "Ausflug." So we all met on the steps of the Institute and walked the short distance out to the country through the woods to a suitable coffee house, talking about algebra, other mathematical topics, and Russia on the way. Evidently this great enthusiasm of Noether's was a major element in her considerable influence on algebraists throughout Germany.

Noether also actively encouraged visitors. I remember well a visit by Emil Artin, then a professor at Hamburg. He had recently been a postdoctoral fellow at Göttingen; he had, with Noether, a considerable influence on the founding of abstract algebra—as represented in the acknowledgments in van der Waerden's book *Moderne Algebra* (1931). During his Göttingen visit, Artin gave three brilliant but condensed lectures on class field theory. I recall meeting Artin subsequently over tea in Noether's or in Courant's home. Artin was, as always, articulate and full of specific promising problems. He formulated to me with emphasis some such explicit problem in class field theory; though it sounded attractive, my knowledge of this theory was then so sketchy that there was no hope of my tackling it. I did not dare admit this ignorance.

Olga Taussky took notes of Artin's three lectures. They were mimeographed and widely used, since they were the most modern presentation of class field theory. Recently they have been translated and published, as an appendix in Harvey Cohn's book on algebraic number theory (1978).

Emmy Noether also was an active host for a visit by the Russian topologist Paul Alexandroff. He gave (in excellent German) lectures on the elementary concepts of algebraic topology, as subsequently published in the Springer booklet of that title (1932). Evidently Noether and Alexandroff talked about mathematics a great deal. It is clear that Noether's enthusiasm for using the appropriate abstract algebraic concepts had a major influence on the development of algebraic topology at that time. Specifically, up until 1926 or later, Alexandroff, Hopf, and others had measured the homological connectivity of topological spaces in terms of numbers: the Betti numbers and torsion coefficients. It was Noether who emphasized the fact that one should replace these numerical invariants by the abelian homology groups of which they are the invariants. The clearest evidence of Noether's early influence is a report [29] of her talk at the Mathematische Gesellschaft in Göttingen on January 27, 1925, where she discussed the introduction of homology groups and observed that the Betti numbers and torsion coefficients were just the standard module-theoretic invariants of these abelian homology groups. This is the earliest example I know of the use of group theory in homology.

This striking example touches only a small part of Noether's wide influence. She developed ideal theory in general rings from the known special cases of number rings and polynomial rings; she understood the use of modules and pioneered the study of cross product algebras; she contributed to relativity theory. In each case, she knew how better to understand mathematical results by using the right conceptual approach. With Artin, she was the leader of the dynamic development of abstract algebra in Germany. [I have tried to give a more complete account of this influence in my article *Origin, Rise, and Decline of Abstract Algebra* (1981).]

Paul Bernays was also an Ausserordentlicher Professor and an assistant to Hilbert. In fact, he was then busily at work on the preparation of the first volume of the monumental Hilbert-Bernays book *Grundlagen der Mathematik* (1934-1939). I knew of his lively interest in logic so I immediately signed up for his course *Elementary Mathematics from the Higher Standpoint.* This course was one which had been started by Felix Klein who felt in particular that it was excellent training for future teachers in high schools (gymnasia, etc.). Evidently, Bernays had been called to carry on this tradition. He did so, but in a fashion which I then found rather dull, with elaborate discussions of different ways of treating the foundations of geometry, carried out in Bernays' careful style, and using his extensive knowledge of the full literature. Bernays was the man who, in effect, directed the preparation of my thesis, which was in logic under the title of *Abbreviated Proofs in the Logic Calculus* (1934). I learned a great deal from him, though in retrospect I do not think I then fully appreciated the extent and precision of his scholarly knowledge. This was also before his splendid sequence of papers on the axiomatization of set theory using both sets and classes (the so-called Gödel-Bernays axiomatics).

There were many other junior faculty members (Privatdozenten, who were entitled to give lectures). For example, I recall learning about partial differential equations from enthusiastic lectures by Hans Lewy. Rellich and others gave informative lectures. There were many visitors: Once each week there was a meeting of the Mathematische Gesellschaft. First came a sit-down tea at a long table and then a lecture in the adjoining room by the visiting mathematician. The professors always sat in the first row, the junior faculty behind them, and the students far in back, but the atmosphere was cordial and there was sometimes a Nachsitzung for dinner in a restaurant. I recall in particular a lively visit from Oswald Veblen whom I met there for the first time, though neither then nor now did I understand his version of projective relativity theory.

One of the best restaurants in town was in the railroad station, not far from the Mathematical Institute. Nearly every noon a number of the Dozenten and senior students repaired there for luncheon. After a bit of introduction I often joined them and found this a lively and stimulating affair. One also met one's fellow students in the well-stocked library of the Institute and between lectures. The usual lecture was for two hours with a fifteen minute break between the hours. In the break, one walked up and down the hall outside, admired the numerous mathematical models displayed in elegant glass cases, and talked to other students.

Göttingen was at that time a major center of mathematical research, setting the style for centers everywhere. Nevertheless, the number of full professors was very small and one wonders in retrospect how this preeminence was achieved. I am unclear as to the answer. Part of it lies in the provision of a considerable number of junior faculty members (Privatdozenten and assistants who were there on temporary appointments). Also the major professors, according to long-standing German tradition, had assistants who gave them a great deal of support and took care of much of the routine work and writing. There must be many other less tangible aspects of Göttingen's dominance. Certainly, its long tradition stretching back to Gauss and Riemann is an important element.

Mathematics at that time was surely not as finely subdivided into specialties as it now is. This must have made it easier for one place to be in the lead in most of the currently important fields—especially when Hilbert and others "took over" promising ideas (like integral equations) which started elsewhere. At any rate, in many of the then prominent specialties, Göttingen was clearly the leading center:

Mathematical logic	Hilbert, Bernays, and others (also Vienna)
Lie groups	Weyl, Herglotz, etc. (also Paris)
Algebra	Noether (also Hamburg)
Algebraic geometry	van der Waerden (also Leipzig)

Analytical number theory	Landau (also Cambridge, England)
Partial differential equations	Hans Lewy and others (also Berlin)
Functional analysis	Rellich and others

In addition to these major fields, Göttingen also had activity in differential geometry (Cohn-Vossen, Fenchel), in the history of mathematics (Otto Neugebauer), and in the philosophy of mathematics (Geiger, a professor of philosophy, occasionally taught courses in this subject).

With these other activities, Göttingen was a heady and exciting place for mathematics. The standards were high. My final oral examination with Herglotz may have been easy, but that with Weyl was hard. At that time, Bernays had been dismissed and Weyl was my sponsor. I seem to recollect that he expected me to know everything. I didn't; I had forgotten the separation axiom for a Hausdorff space.

In the spring of 1933 it all came apart. Hitler and the National Socialists won two elections in rapid succession in January and February of 1933 and Hitler became Reichskanzler. Among other things, this gave him power over the universities, which were all state universities. Very soon a decree was issued dismissing, as of the beginning of the spring semester, all those faculty members who were wholly or partly Jewish. Exception was made only for some of those, like Geiger, who had served as soldiers in the First World War, but this exception lasted only a semester and left these professors in a very troublesome position. For example, that spring I took Geiger's course on the philosophy of mathematics; I could not help noticing his constant distress. Courant, who had served in the First World War, was nevertheless dismissed immediately from his position as director of the mathematical institute. He was succeeded by Neugebauer who lasted exactly one day as director. The rumor was that he would not take orders. The institute did continue to operate during the spring semester, but under evident conditions of strain with many faculty searching for suitable positions in other countries and many students hurrying to get theses done. By the following semester over half the people I have named were elsewhere and the great days of Göttingen were fatally interrupted.

APPENDIX A: MATHEMATICIANS AT GÖTTINGEN 1931-1933

Ordentliche Professoren (Full Professors)

David Hilbert (emeritus)
Edmund Landau
Gustav Herglotz
Richard Courant
Hermann Weyl

Ausserordentliche Professoren (Associate Professors)

Paul Bernays
Paul Hertz
Emmy Noether

Privatdozenten

Wilhelm Cauer
Hans Lewy
Otto Neugebauer
Arnold Schmidt
Werner Weber
Udo Wegner

Assistenten

Herbert Busemann
Max Deuring
Werner Fenchel
Heinrich Heesch
Hans Heilbronn
Rudolf Lüneburg (mathematical optics)
Wilhelm Magnus
Kurt Mahler
E. J. McShane
Franz Rellich (P.D.E.)
Olga Taussky
Helmut Ulm
Stefan Warschawski

Studenten

Gerhard Gentzen
Charlotte John
Fritz John

Gerhard Lyra
Meyer-Leibniz (student of physics)
Peter Scherk
Arnold Schmidt
Kurt Schütte
Hans Schwerdtfeger
Oswald Teichmüller
Ernst Witt

Occasional Visitors

Paul Alexandroff (Moscow)
Emil Artin (Hamburg)
Reinhold Baer (Freiburg)
Kurt Friedrichs (Brunswick)
Heinz Hopf (Berlin, Zürich)
Ruth Moufang (Frankfurt)
Alexander Ostrowski (Basel)
George Polya (Zürich)
B. L. van der Waerden (Leipzig)
Oswald Veblen (Princeton)
Richard von Mises (Berlin)
John von Neumann (Berlin)

APPENDIX B: A VIEW OF THE LECTURES IN MATHEMATICS AT GÖTTINGEN

As a contemporary description of mathematics courses in the fall semester of 1931, I transcribe here a letter which I wrote my mother on December 8, 1931, describing these lectures.

Undoubtedly the most beautiful and well-arranged lectures which I hear are those of Herglotz on Continuous Groups. Herglotz is noted for the clarity of his lectures; he prepares them well and delivers them with great energy and with great attention to pedagogical detail. One can always understand him without bothering to do any outside studying on the subject—this of course has its disadvantages; one doesn't digest the subject so thoroughly.

The lectures by Bernays are particularly interesting for me, because they treat the foundations of mathematics. Hitherto he has considered the foundations for ordinary (Euclidean) Geometry, but soon he will take up the discussion of Algebra. The lectures are perhaps not as polished as those of Herglotz, but they cover a great deal of ground and are very suggestive.

The seminar with Prof. Weyl is also very good. It comes but once a week; but every time before the class we have a social gathering with tea and cookies. In the seminar itself various students give reports on topics taken from recent and pertinent mathematical literature. Of course Prof. Weyl continually makes comments and suggestions. He also is very energetic; his methods of thought are far-reaching and inclusive. I am sure I can learn a great deal from him.

Hilbert's lectures—once a week—are very heavily attended; one has to go early in order to get a good seat. On the first day the original room was not large enough, and a larger room also did not suffice, so that a number of students had to stand. But, as usual, the number of the hearers declines as the semester progresses. Hilbert is now quite old, and so can't go so rapidly—in fact he has an assistant to write symbols on the board for him. Nevertheless the lectures are very interesting as an integration of various elements of modern science.

Prof. Noether's lectures (she is a woman—a member of a noted mathematical family) are also excellent, both in themselves and because they bear an entirely different character in this excellence. Prof. Noether thinks fast and talks faster. As one listens, one must also think fast—and that is always excellent training. Furthermore, thinking fast is one of the joys of mathematics. Furthermore, the subject of the lectures is very closely allied with the thesis I wrote in Chicago last year.

REFERENCES

Alexandroff, P. (1932). *Einfachste Grundbegriffe der Topologie,* Springer, Berlin.
Cohn, H. (1978). *Algebraic Numbers,* Springer, Berlin.
Hilbert, D., and P. Bernays (1934-1939). *Grundlagen der Mathematik,* Springer, Berlin.
Landau, E. (1932). Uber die Fareyreihe und die Riemannsche Vermutung, *Nachr. d. Gesselch. d. Wiss. Zu. Göttingen,* 347.
Mac Lane, S. (1934). Abbreviated Proofs in the Logic Calculus, *Bull. Amer. Math. Soc. 40,* 37-38.
Mac Lane, S. (1981). Origin, Rise and Decline of Abstract Algebra, in *American Mathematical Heritage: Algebra and Applied Mathematics,* Graduate Studies, Texas Tech University.
Van der Waerden, B. L. (1931). *Moderne Algebra,* 2 volumes, Springer, Berlin.

3

My Personal Recollections of Emmy Noether

Olga Taussky

In my high school days the name of Sonja Kowalewsky was already a household word. Her biography, written by the sister of Mittag-Leffler, had been published in the inexpensive Reklam series. Other women mathematicians seemed known only to experts. However, when I became a student at the University of Vienna (Wien) in Austria the name of Emmy Noether emerged for me too. My teacher, Philip Furtwängler, devoted some time to her work in Galois theory; the amusing story of how Hilbert fought for her appointment to the University in Göttingen was told; people visiting Germany told of having met her; Furtwängler's student Gröbner called on her in connection with his thesis in algebraic geometry; and there were other connections which made me realize her growing importance in the mathematical world. Most of her work was not yet meaningful to me. (For my training was in number theory, the kind of number theory which involves numbers; this is also my favorite subject.) By the time I met Emmy in Königsberg at the meeting of the Deutsche Mathematikervereinigung in 1930, she had started her own work in number theory and gave deeper insight into some of the results of algebraic number theory. This was done through her methods of abstract algebra. I myself gave a short paper at this meeting. It concerned my thesis, written under Furtwängler, entitled *Über eine Verschärfung des Hauptidealsatzes.* As soon as I had finished Emmy jumped up and made a quite lengthy comment which, unfortunately, I was unable to understand because of insufficient training. However, Hasse understood it and replied to it at some length, and there developed between these two mathematicians some sort of duet which they enjoyed thoroughly. Clearly, Emmy was pleased and I even overheard some nice remarks she made about my lecture. She spoke to me frequently later, but not about the subject of my talk. She was very friendly; so was Hasse. All this was very helpful to me, for prior to my lecture I had been justifiedly very nervous, for this was a meeting to which very famous mathematicians came, including even Hilbert. Königsberg was Hilbert's home town and he delivered there one of his most famous speeches (a recording of which is available).

When lunch came I sat down next to Emmy, to her left. I have the following recollection of this lunch. Emmy was very busy discussing mathematics with the man on her right and several people across the table. She was having a very good time. She ate her lunch, but gesticulated violently when eating. This kept her left hand busy too, for she spilled her food constantly and wiped it off from her dress, completely unperturbed. Life was still quite peaceful these days and Emmy was having a great time. She worked a great deal with Hasse then.

The meeting played a great role in my mathematical life, for I met A. Scholz. I could have met him in Vienna when he came there in connection with his thesis (written under I. Schur). But I was still a student then and there was no occasion to meet. His work was also in class field theory, like my own, and there were strong connections between the subjects we were interested in. At Königsberg he was lecturing on logic. We decided to start on an investigation of the cubic class fields of imaginary quadratic fields, both from the number-theoretic and group-theoretic (via Artin's newly introduced nonabelian methods) point of view. The completion of this paper took quite a long time, for we met only at mathematical conferences and progress was slow.

The next time I met Emmy was on a similar occasion, a year later. But there we had very little contact. I recall her being very kind and helpful to Zermelo. I again gave a paper in class field theory, worked with Scholz, met his sister with whom I started a very pleasant friendship, but above all, through the help of one of my Vienna teachers, Hans Hahn (of Hahn-Banach fame) made the personal acquaintance of Courant, who had already heard of my lecture. All this was lucky, for in those days of almost total unemployment for young people, Courant was much in need of a young person trained in class field theory to help with the editing of Hilbert's collected papers in number theory.

Shortly after my return from the meeting I received in fact an invitation to come to Göttingen and I again met Emmy there. She immediately announced to me proudly that she and Deuring, her favorite student, had studied class field theory in the meantime and that she was going to run a seminar on this subject because of my visit. So I had two important reasons for contact with Emmy during that year. In the seminar various parts of class field theory were discussed. I lectured on an extension of Hilbert's Theorem 94 to cyclic unramified extension fields of degree p^r instead of p. Once the proof of Furtwängler of the Hauptidealsatz came up and Emmy repeated what almost everybody said, namely, that it was an unattractive proof. I came violently to Furtwängler's defense. It seemed unfair to criticize the first proof of a theorem that had been a challenge to number theoreticians for decades, ever since Hilbert had formulated it. After all, it may not have been correct! I was amazed by my daring, and so were the others. However, Emmy was completely calm, and I feel certain, was not in the least angry with me. I recognized for the first time that she was a person who did not mind criticism.

Probably I will always be puzzled as to why the mathematical world harped on the ugliness of the proof far more than on the result it achieved. Nobody will deny that Iyanaga's proof that appeared later is incomparably more pleasing and revealing. Furtwängler used Schreier's treatment of group extensions and some sort of group ring calculus which enabled him to carry out his computations. Schreier was Furtwängler's student, and Furtwängler must have developed these tools over quite some time. I myself owe to this method a good bit of my training in p-groups and group extensions. It helped me later to understand Lyndon when he told me about his thesis, which involves cohomology, in which his teacher S. Mac Lane played such a decisive role.

Emmy took some interest in my work as editor of the Hilbert volume in number theory. She herself had been an editor of Dedekind's three volumes, and this work had given her great satisfaction. She came to appreciate Dedekind's work to the utmost, and found many sources of later achievements already in Dedekind. Occasionally she annoyed even her friends by this attitude. She managed to rename the Hilbert subgroups the Hilbert-Dedekind subgroups.

I was asked by the *Monatshefte* to write a review of the Dedekind volumes. This was quite a worry to me, but I felt very pleased when the publisher (Vieweg) quoted a sentence of my review in their advertisements, with my name attached. I suppose they did not realize what a beginner I still was.

Emmy was truly amazed when I told her that Hilbert's work contained many errors. She said that Dedekind never made any errors. Hilbert's errors were on all levels. The three editors, W. Magnus, H. Ulm, and I worked exceedingly hard, but were greatly handicapped by lack of time. The first volume, on number theory, was to be completed for the celebration of Hilbert's 70th birthday in January 1932. The publisher of the volume, Springer, was to come to Göttingen to present the volume, bound in white leather, to Hilbert. Unfortunately, this was totally unrealistic and it was not the finished product that Springer handed over. The fact that the work contained errors had come as a surprise. Actually, we did not find all the wrong statements, but owing to my training I was able to point to conjectures which had turned out to be incorrect. False rumors had spread that our work concerned small defects only and delayed the publication unnecessarily, and this was even mentioned in C. Reid's volume on Hilbert.

But we had to endure even worse trouble during our task, and there Emmy was somehow involved too. The preparation of this volume had become known. I myself informed the few people who were my mathematical acquaintances at this early time of my life. These included Fueter and Speiser, then still in Zurich, where I had spent a semester when still a student after my thesis was already completed. They sent me a list of minor corrections which I was happy to use, but they seemed to be asking for more; they did not like the *Zahlbericht* and thought this was a good time to rewrite it. Emmy had been traveling during Christmas vacations and she heard similar sentiments from other places. Also I. Schur told her there were some errors. All I could do was write to all the people who complained, but, of course, there was no question of rewriting the *Zahlbericht*. Emmy herself never suggested it. Hilbert's *Zahlbericht*, which is included in Vol. 1 of the Collected Works, is really a book, but was published in the *Jahresbericht* of the Deutsche Mathematikervereinigung (DMV). It is probably not as well known now as in my student days. It deals with the theory of algebraic numbers, from the most elementary facts to some of the important results in class field theory. It includes Theorem 90 (a term introduced by Artin) and the much deeper Theorem 94, a term introduced by myself, and a great deal more. The only other really well-known text on algebraic numbers in German at that time was Hecke's book. Hilbert's great articles on class field theory, containing his important conjectures, were published in two separate papers. It was only later, at Bryn Mawr, that Emmy burst out against the *Zahlbericht*, quoting also Artin as having said that it delayed the development of algebraic number theory by decades. I must admit that at that stage I did not even understand what these complainers meant. It was only through my own work on integral matrices that I realized where one of the troubles was—that being trained mainly in work in the maximal order does not get one very far, and also that methods of abstract algebra and local treatments are most helpful and revealing. It is well known that Emmy's influence in these directions was supreme.

The maximal order of an algebraic number field, as it is called nowadays, is its ring of integers. It is a Dedekind ring and many of its fundamental properties can be obtained more elegantly by treating it just as a Dedekind ring. However, the suborders of integers play a very big role too. There the ideal classes do not form a group any longer, for they need not have inverses. E.g., an ideal in such an order can satisfy the equation $a^2 \sim a$, without $a \sim 1$. These suborders are Noether domains, as they are called now. They obey other axioms and were studied intensively by Dedekind, who showed among other things that the number of ideal classes is still finite.

I had very little training in abstract algebra from Furtwängler; the books of van der Waerden had hardly appeared. Furtwängler was most powerful in number theory and group theory, but behind the times in the abstract algebra trend which was already flourishing in Germany, the USA, and other places—in Germany mainly through Artin, Schur, Emmy, van der Waerden, and in the USA mainly Dickson and Wedderburn. In addition, Furtwängler's physical disability impeded his contacts.

The Hilbert work kept me exceedingly busy. Apart from this I was also asked to grade for Courant's course on differential equations. So I was hardly able to work on my own problems, nor profit from the abundance of mathematical wisdom of the place. I attended Emmy's course on representation theory, but found it hard to follow. I recall that this course was also attended by the great number theoretician K. Mahler and by E. Witt, who was still a student then but had already made a name for himself with a new proof for Wedderburn's theorem on finite fields. He became Emmy's student and did very distinguished work later.

Emmy herself had had a great year. She had visits by Hasse and by van der Waerden. She had her students, the so-called Noether boys, and was a great success as a teacher with them. Let me tell a story about them. During one of her lectures, on December 6 (St. Nikolaus day in Germany and Austria), the door of the lecture room opened and a giant Lebkuchen was brought in iced with an inscription concerning Emmy's students. Everybody was startled. Emmy asked in amazement, "Who sent this?" Somebody said, "Perhaps Hasse." This seemed possible, because he was singing praises of her proof of the Hauptgeschlechtssatz, the principal genus theorem. So the Lebkuchen was cut into pieces and one of the boys suggested sending a thank-you note to Hasse. This was done with all sorts of jokes, e.g., from the graders' room a stamp "Womit nichts bewiesen ist" was fetched. Emmy wrote, "Fräulein Taussky schafft noch immer," referring to my slow eating. However, it was not Hasse, but the Noether boys who provided the cake!

As is known, Emmy did not have the position of Ordentlicher Professor, not everybody liked her, and not everybody trusted that her achievements were what they later were accepted to be. She irritated people by bragging about them. This irritation extended even to Deuring, although he appeared to be a shy, quiet man. I remember a professor asking me whether he really had achieved that great work on the Riemann hypothesis that later led to Heilbronn's proof of Gauss' conjecture on the class number in imaginary quadratic fields. He wondered whether it was just another exaggeration of Emmy's. I assured him that this was not the case. One day I was present when one of the senior professors talked very roughly to her. The next day I told him (disregarding his influential position) that this had upset me very much. Apparently "big shots" are not necessarily mean, for this one went to apologize to her. He even repeated his apology to my colleagues, assuming that I had told them of his insult—but I had not!

Outside of Göttingen, Emmy was greatly appreciated in her country. I overheard Professor Krull saying, "Miss Noether is not only a great mathematician, she is also a great German woman!"

Emmy had her 50th birthday in 1932 and told me about it. She commented that nobody at Göttingen had taken notice of it—at that time all birthdates were published in the *Jahresbericht*. But then she added, "I suppose it is a sign that 50 does not mean old."

Emmy had cause to have grievances against some of the people at Göttingen. But at the time I knew her at Göttingen these grievances were not very important to her. It was the academic year 1931-1932 and she was at the height of her power and proud of her achievements, knowing that her ideas were now being accepted. (I have a foggy memory of her saying this to me—I had felt proud that she accepted me sufficiently at that early time of my life to communicate such feelings to me.) Others will review in this book her main achievements, describe her philosophy of mathematical proofs, and her enormous and lasting influence on her students and on mathematics. But during the year 1931-1932—and I checked this in her bibliography—three important publications of hers appeared: [38],* in the Hensel *Festschrift* of Crelle's Journal; [39], with R. Brauer and H. Hasse; and [40], the lecture at the Zurich Congress. I do not think that I was directly influenced by these papers when they appeared. But later in my own very different domain of research all three subjects treated in them play a role. I especially recall two papers by her student Deuring (1932, 1931) of that same period. Emmy seemed particularly interested in the first of these papers.

*Numbered references are to Noether's papers listed at the end of this book.

Fröhlich's contribution to this book deals with Emmy's influence on algebraic number theory—class field theory, with cohomology playing a vital role. It seems certain that she could have done more there. It is futile to wonder what it might have been.

Emmy and I were very friendly until the end of the academic year, continuing through the International Congress in Zurich in 1932. She talked to me about the lecture she prepared for this occasion. Realizing that only the great experts would understand what she had to say, I suggested that she give some simple examples at the start, and she actually did this. The lecture is entitled *Hyperkomplexe Systeme and ihre Beziehungen zur kommutativen Algebra and Zahlentheorie* [40] (see Chapter 11 of this book).

In this lecture she stressed the importance of noncommutative methods in the study of problems concerning commutative systems. How I wish I could show her my own attempts in this direction, mainly via matrix theory. In particular, the following fact would have amused her: Let A, B be 2×2 rational matrices, where A, say, has an irreducible characteristic polynomial. Then $-\det(AB - BA)$ = norm from field of eigenvalues of A. Of course, she would have seen the link with cyclic algebras at once.

During 1932 Artin came to Göttingen to give three lectures on the current state of class field theory. People from other parts of Germany came for this event. I took notes on these lectures, which were mimeographed and widely distributed and used. Recently they were translated and published [Appendix 11 by O. Taussky in the book of Cohn (1978)].

On the occasion of Artin's visit I met his young wife Natascha, born in Russia, for the first time. She was very friendly to me. We suddenly realized that we had both been in the same high school in Linz, Austria, but not in the same classes. There had indeed been a Russian girl there, wearing knee stockings on the coldest winter day and taking private lessons in German.

I met van der Waerden for the first time in Göttingen. I soon found him a most stimulating person whose ideas had great influence on my work. I recall Emmy asking him how it came about that he could write a yellow book on quantum theory. In recent years he has reproved a theorem of mine by the use of invariant theory (Taussky, to appear). I am sure Emmy would have appreciated this.

I did not return to Göttingen afterwards, but recall only one more connection with Emmy. She was referee of a paper of mine for the *Mathematische Annalen*. She made some comments and accepted it.

After I left Göttingen in July 1932 I went back to Vienna and spent the years 1932-1934 there helping in the mathematics department, partly on a voluntary basis and partly with small assistantships, and earning money through private tutoring. I continued my own research, of course, but was rather isolated. However, all of a sudden I received an invitation to go to Bryn Mawr College in 1933-1934. This was arranged through Veblen, whom I had met in Göttingen and who was acquainted with the head of the department at Bryn Mawr, Mrs. A. Pell Wheeler.

However, my visit was canceled because of financial losses suffered by the college during the depression. In the meantime Emmy arrived there, a visit arranged by the Rockefeller Foundation. She had a good time there, studying van der Waerden's Vol. 1 with the staff and a small number of students. In 1934 the college was able to invite me too for the academic year 1934-1935, an invitation now much more profitable because of Emmy's presence. I was to receive a scholarship for foreign students, a position rather poor from the financial point of view. However, a short time after I had accepted, I was informed that Girton College, Cambridge, had awarded me a three-year research fellowship. Because of Emmy's presence at Bryn Mawr they allowed me to proceed to Bryn Mawr for one year. So after several hardships in my career things were looking up again. I was busy with various duties to the last moment, took some hurried lessons in English, and then took off, arriving at Bryn Mawr in late September in the afternoon. In the evening, one of the graduate students, Ruth Stauffer, took me to Emmy. Now, Emmy was not in the best mood. This was clear to me from the first moment I saw her. She had, in fact, a considerable number of troubles on her mind. She did not confide this to others. Knowing the end of the story, her death in early spring, one can understand what worries and anxieties she carried with her. Life is not so very different from a detective story. When you read the end you realize how many subtle hints were given to you to guess the outcome, but you ignored them and concentrated on obvious facts.

After her pleasant time in 1933-1934 she had gone back to Germany in the summer of 1934. There she realized that the political situation had deteriorated immensely. Some of her colleagues even avoided her there. She decided to move her belongings out of the country, to an uncertain future. She was not at an easy age for a woman and in any case in those days her age was considered rather advanced, particularly for a woman. She had no position for the next year, and, in fact, still had none at the time of her death. She knew that her friends would do something for her, but they seemed very slow to do so. She felt certain that she would not accept a position to teach undergraduate students at Bryn Mawr. I do not know when she became aware of health problems, but later it seemed to me clear that there were days when some utterances of hers referred to it. She had apparently hoped to spend the next summer in Germany again and have her doctor repeat an operation he had performed previously. But her condition worsened in the spring.

I do not wish to give the impression that she was in a bad or depressed mood during all that academic year. But she certainly was in a very changeable mood. However, I now had far more contact with her than in my Göttingen days and so cannot really tell how much she had changed. She also seemed changed in her reactions towards me. Now she was definitely moody; I could never predict how she would react to anything I did or said. She seemed very pleased to have somebody with whom she could converse in German—even if she did not care for my Austrian accent. She felt close enough to me to criticize a great many personal things about me, and, on the other hand, to scold me for being weak if I yielded to her criticism.

Let me give three instances. When I arrived at Bryn Mawr I was the proud owner of a green Austrian felt hat with a feather. Women wore hats then. Emmy nagged me about this hat every time she saw me with it. She said I reminded her of a certain mutual acquaintance when he wore Lederhosen. So I gave the hat to Ruth Stauffer who I knew loved it.

Second, I used to wear small scarves, knotted over in front, while the other girls at Bryn Mawr knotted theirs behind their necks. She said I looked like a "Berlin Droschkenkutscher."

Third, I communicated to Emmy that I had heard about a store in New York where one could buy European chocolate bars. She was very interested in this because she owed a thank-you gift to the Brauers for hospitality in Princeton. So from my next expedition to New York I came back with a few bars. However, they were milk chocolate, while Emmy preferred bitter ones. She gave them to the Brauers with deep apologies, but they claimed that the creamy variety was more to their liking and she was very pleased. She was proud of me and wanted me to make a good impression, but on the other hand ridiculed me in the presence of others. I seemed to be the only person on whom she could let her temper go, and now, when looking back, I feel I ought to be proud of this. I was not always proud of it then, because I was anxious for the sake of my own career. In some respects she was like a moody child. But she could also be motherly, understanding, and charitable.

The best time both of us had together was when I accompanied her to Princeton where she gave a lecture every Tuesday, traveling there early morning and returning late in the evening. I did not always accompany her because of the expensive fare. But when I did she seemed to enjoy my company and we talked in the train and while waiting at Princeton Junction or in Philadelphia. We discussed work and many other things. My trips to Princeton were the highlights of that year.

Emmy held a contract with the Institute of Advanced Study for a two-hour lecture per week. During the two years since I last saw her I had changed a great deal. In spite of my isolation in Vienna I had learned a great deal of mathematics and had grown in mathematical insight. While still in Göttingen I had not discussed the problems of my thesis with her at great length, but now I talked more about them, including my now finished paper with Scholz. In the meantime she had also met Scholz and discussed this kind of work with him. He was one of the very few who had not avoided her. It amazed me that he was able to make her listen to numerical details and even make her appreciate them. Later, during my stay, she even computed a tiny example on the same subjects. In typing the joint paper I suddenly added another paragraph and while I did this the idea of the group tower suddenly came to me. This was the chain of Galois groups of the class fields in the class field tower. Surely, if one could prove that no such tower of infinite length could exist, then the class field tower could not be infinite either. The work of Scholz and myself, apart from the easy examples of cyclic groups and the four group, seemed to indicate such a behavior. Scholz did not really go with my theory entirely, nor did Emmy. Further Magnus, my former colleague from the Hilbert volume, showed that group towers of arbitrary length do indeed exist. Later, Ito, Hobby-Zassenhaus, and finally Serre showed that for all types of class groups infinite group towers can be constructed, and Golod and Safarevic then went ahead and constructed a numerical example with infinite class field tower.

Magnus was in Princeton during that same year, so that he, Emmy, and I were a little remnant of Göttingen in 1932. She forwarded the Magnus paper to the *Annalen*. She was closely attached to this journal, but not one of the editors; I understood that this was a disappointment to her.

In Princeton I also met Weyl again; of course, Veblen; W. Mayer, the assistant to Einstein, who had been in Vienna previously; R. Brauer, whom I had met in Königsberg and who looked after Emmy and myself in many ways; and von Neumann, whom I had met at Menger's Kolloquium sessions in Vienna and later in Göttingen when he delivered a lecture on his newly found result concerning compact groups. Many of these scholars entertained Emmy, and since I traveled with her they asked me too. The Institute was still housed at Fine Hall and seemed to me a wonderful place. One morning Morgan Ward introduced himself to Emmy and asked if he could discuss his work on recurring sequences with her. Later I was to meet him at Caltech at a number theory conference in 1955. He played a role in my appointment as the first woman to teach here. However, he had spread the rumor that I was Emmy's student and so they suggested to me to teach the advanced algebra course right away, while I was actually at that time occupied with very different things. H. F. Bohnenblust and H. P. Robertson were at Princeton then. Bohnenblust was chairman at Caltech when I was appointed there and Robertson was Professor of Mathematical Physics. He also supported my appointment.

On one of the first trips to Princeton I told Emmy on the train that while still in Vienna I had become interested in topological algebra. This came about through Pontryagin's paper *Stetige Körper* (1932). Topological algebra was in the air in Princeton too and it was mainly abelian groups. I realized that considering a field as an abelian additive group would lead to Pontryagin's results. Emmy was very helpful about this. She introduced me to Alexander, and he said that Jacobson was an expert on Pontryagin's paper and that I ought to discuss this matter with him. This led to a joint publication, *Locally compact rings* (Jacobson and Taussky 1935).

Emmy's Princeton lectures Tuesdays were a repeat of what she worked out with us at Bryn Mawr on Mondays. She somehow used us as a rehearsal. I remember that on one occasion she had to apply the binomial theorem to a very special situation. Although she had enormous insight into difficult abstract structures this computation was a great challenge to her. She did however master it and was very pleased about this. In fact she turned back from the blackboard about three times to smile proudly at us. We were quite a small group. There was her Ph.D. student Ruth Stauffer (McKee) who worked with her on normal bases in fields and was her favorite. Emmy always associated her with a woman called Stauffer in Schiller's play *Wilhelm Tell* and decided that she must somehow be of Swiss ancestry. The Stauffer thesis (1936) is connected with previous work by Speiser. There was Grace Shover (Quinn) who had been trained in abstract algebra by MacDuffee and had worked on ideals and class numbers in algebras. Then there was the oldest and most settled of us, the late Marie Weiss, a student of Manning who had worked on multiply transitive permutation groups. To her Emmy gave a definite problem which led to a publication later (Weiss 1936) and was concerned with units of normal fields. This work was connected with previous work by Latimer and it pleases me personally that Emmy appreciated the attempts by Latimer and also by MacDuffee. These two algebraists did work that I found later on to be stimulating for my own enterprises.

The head of the department at Bryn Mawr was Anna Pell Wheeler, a very dignified and distinguished lady. She seemed the most important woman mathematician in the USA at that time. Her subject was functional analysis and she had given a course of AMS Colloquium lectures. She also lectured to us in 1934-1935 on this subject. A twice-widowed, elegant lady, she had a different personal background from Emmy. She was much interested in teaching, even undergraduate teaching. In spite of the differences there seemed to be real friendship between the two women. They were of about the same age. Mrs. Wheeler, who was supposed to have been a tower of strength previously, seemed rather frail and unwell at the time I met her, but Emmy did not take her sufferings too seriously, nor did she approve of me when I declared myself ill once or twice that year. I think perhaps she had a point there.

Emmy considered herself a "tough guy" who would not yield to sickness. She did not take her surgery too seriously and did not expect the sad outcome. She was looking forward to a somewhat slimmer body.

In Emmy's seminar on Mondays her main plan was to study the mimeographed lecture notes of Hasse on class field theory. [They were published later (Hasse 1967).] However, these notes were not meant for people who hardly knew any algebraic number theory. She made me give some introductory lectures on this subject (and repeat parts of this in Princeton when I went with her). However, it now emerged that my training in the fundamentals of this subject was like it was in Hilbert's *Zahlbericht* and—as I said earlier here—Emmy, who had kept quiet at Göttingen about her dislike for this, now flared up. So this was a tough situation! She called my approach not algebraic, but "Rechnen," i.e., computing! While the van der Waerden books enabled the world to learn abstract algebra, nothing similar was yet available for algebraic number theory. It took decades for books to appear to fill this gap. However, some of these books contain no numbers at all and this is not the right remedy either.

I had not had, either in Göttingen or Bryn Mawr, any systematic instruction from Emmy on algebra. It was more a question of picking up some items. One of them is the factor sets, a forerunner of cohomology. I enjoyed this concept so much that I worked out examples. When she saw me doing this for several days she laughingly persuaded me to stop this game. The term "Noether equations" connected with this was introduced later.

Of course, teaching did not absorb all of Emmy's time at Bryn Mawr, but I do not have much knowledge of her other activities. I know that she was busy giving Deuring advice on his *Ergebnisse* volume entitled *Algebren* (1935); at one time she suggested that some of Scholz's and my ideas might be mentioned, but it was already too late for that. I know that she continued her correspondence with Hasse. Of course, people came to consult her, but she was not likely to have discussed this with me.

Emmy was not uninterested in the many problems women face. I know that she was concerned already in Göttingen. I think it was through her, but am not completely certain about it, that I learned about the IFUW, the International Federation of University Women, of which the AAUW is a branch. I recall that at the congress in Zurich in 1932 she attended one of their meetings when they invited her, or maybe she only mentioned the invitation to me. In any case I do recall that she said that one ought to attend such functions. Later my affiliation with this federation was very helpful to me. She was very interested in the girls at Bryn Mawr. She did not approve of girls living in such an isolated place and would undoubtedly have been in favor of present-day arrangements.

She was in favor of career women marrying (though, I think, she did not give a moment of her time to think of the practical problems involved in this). In fact she and Ilse Brauer, Richard Brauer's wife, tried to marry all four of us off, but did not succeed.

But for herself she was content with the way things had turned out and did not resent her early struggles. In fact she now felt that women ought not to arrange their duties the way men did. Of course, again she did not for a moment appreciate the fact that not all women have the means for such a life style. She was very naive and knew very little about life. She saw women as being protected by their families and even admitted to me that she gave young men preference in her recommendations for jobs so they could start a family. She asked me to understand this, but, of course, I did not. Then she suddenly burst out and attacked me for the "expensive" dress I had worn at Hilbert's 70th birthday party. But it had been inexpensive, bought at one of the lesser department stores in Vienna. (This dress has even made its way into Constance Reid's book!) I got rather upset and worried, but since she was really a kind person and also started to understand my feelings she apologized profoundly.

Looking back I think that Emmy had warm and appreciative feelings towards me from the start, and this was confirmed to me by Alexandroff whom I met many years later at one of the international mathematical congresses. He was probably the closest co-worker and admirer she had.

Before she left her home for her fatal operation she was persuaded to make a list of people to whom she would like to be given an item of her belongings. She included me in this list, as I was told a good bit later. The lawyers sent me a brooch and a book. She had worn that brooch daily, but actually it was one I had given to her! The book was written by Dedekind and was in fact his personal copy!

I want to close with a fact totally independent of what I have tried to report here. It concerns K. Shoda, a Japanese algebraist of great accomplishments. She had apparently stimulated him very much during his years in Germany and so had I. Schur. His thesis adviser was the famous Takagi. In the Japanese translation of A. Dick's book on Emmy he calls her his "Lehrerin." He did work not only in abstract algebra, but also in matrix theory, and there is a theorem of his I absolutely adore. It states that a matrix with trace 0 is an additive commutator and—apart from a very small number of exceptions—if of determinant 1, a multiplicative commutator of matrices with elements in the field of entries of the given matrix. He did not prove this for all fields. Although I immediately had a great interest in this theorem it took quite a while before I found an application for it. Then followed a paper by Albert and Muckenhoupt proving the additive case for all fields with no exceptions. The multiplicative part formed the important thesis of R. C. Thompson and is widely known.

On the desk in his study in Tokyo, Shoda had photographs of Takagi, Schur, and Emmy!

REFERENCES

Cohn, H. (1978). *Algebraic Numbers,* Springer, Berlin.
Deuring, M. (1931). Zur Theorie der Normen relativzyklischer Körper, *Nachr. d. Gesellsch. d. Wiss. zu Göttingen,* 199-200.
——— (1932). Galoissche Theorie und Darstellungstheorie, *Math. Ann. 107,* 140-144.
——— (1935). *Algebren,* Ergebnisse der Mathematik und ihrer Grenzgebiete, Springer, Berlin. Also Chelsea, New York, 1948, and Springer, Berlin, 1968.
Hasse, H. (1967). *Vorlesungen über Klassenkörpertheorie,* Thesaurus Mathematicae, Band 6, Physica-Verlag, Würzburg. [Ed. note: Noether reviewed the original mimeographed notes in *Zentralblatt 6* (1933), 390.]
Jacobson, N., and O. Taussky (1935). Locally compact rings, *Proc. Nat. Acad. Sci. U.S.A. 21,* 106-108.
Pontryagin, L. S. (1932). Über stetige algebraische Körper, *Ann. of Math.* (ser. 2) *3,* 163-174.
Stauffer, R. (1936). The construction of a normal basis in a separable normal extension field, *Amer. J. Math. 58,* 585-597.
Taussky, O., ed. *Seminar Notes on Ternary Forms and Norms,* Marcel Dekker, New York, to appear.
Weiss, M. (1936). Fundamental systems of units in normal fields, *Bull. Amer. Math. Soc. 42,* 36.

4

Obituary of Emmy Noether

B. L. van der Waerden

Fate has tragically taken from our science a very important, entirely unique personality. Our faithful journal co-worker Emmy Noether died on April 14, 1935 as a result of an operation. She was born in Erlangen on March 23, 1882, the daughter of the well-known mathematician Max Noether.

Her absolute, incomparable uniqueness cannot be explained by her outward appearance only, however characteristic this undoubtedly was. Her individuality is also by no means exclusively a consequence of the fact that she was an extremely talented mathematician, but lies in the whole structure of her creative personality, in the style of her thoughts, and the goal of her will. For as these thoughts were primarily mathematical thoughts and the will primarily intent on scientific recognition, so must we first analyze her mathematical accomplishments if we want to understand her personality at all.

One could formulate the maxim by which Emmy Noether always let herself be guided as follows: *All relations between numbers, functions, and operations become clear, generalizable, and truly fruitful only when they are separated from their particular objects and reduced to general concepts.* For her this guiding principle was by no means a result of her experience with the importance of scientific methods, but an a priori fundamental principle of her thoughts. She could conceive and assimilate no theorem or proof before it had been abstracted and thus made clear in her

Translated by Christina M. Mynhardt, Department of Mathematics, University of South Africa, Pretoria, South Africa, with the permission of Springer-Verlag, from *Mathematische Annalen 111* (1935), 469-476.

mind. She could think only in concepts, not in formulas, and this is exactly where her strength lay. In this way she was forced by her own nature to discover those concepts that were suitable to serve as bases of mathematical theories.

For her, algebra and arithmetic were ideal as material for this way of thinking. She considered the concepts of field, ring, ideal, module, residue class, and isomorphism as fundamental. The model for her abstract development, however, she found primarily in the Dedekind module theory, from which she could always draw new ideas and methods, and whose field of application she markedly extended in several directions.

She started with Gordan's invariant theory. Her dissertation [4; see also Ref. 3]* with which she graduated in 1907 in Erlangen deals with the problem of extending to the n-ary case the methods developed by Gordan for the binary and ternary cases.† She later gave still nicer applications of the n-ary series expansions of invariant theory [8, 16].

Very soon, however, she came under the spell of Hilbert's methods and questions. Her finiteness proofs for invariants of finite groups and for integral invariants of binary forms belong to the Hilbert circle of problems. Her most important work from this period is that on fields and systems of rational functions [6; see also Ref. 5] in which she proved the existence of a finite rational basis for each system of rational functions in n variables by combining the methods of Hilbert's finiteness arguments with those of Steinitz' field theory. By using this result she solved one part of the Hilbert problem of relative integral functions. With the methods of the same work [6] she also made a substantial contribution—the most important that has been obtained hitherto—to the problem of constructing equations with a given group [11].

During the war Emmy Noether came to Göttingen, where she qualified to lecture in 1919 and soon afterwards received an appointment as Professor. Under the influence of Klein and Hilbert, both of whom during this time were very much occupied with general relativity theory, her work on differential invariants [12, 13], which has become very important in this field, was accomplished. Here, for the first time, she gave the general methods suitable for generating simultaneous differential invariants. In the first work the fundamental concept of reduction system is established: a system of differential invariants of which all the others are algebraic invariants. In the second

---

*The numbered references are to Noether's papers listed at the end of this book.
†Editor's note: Actually, the dissertation was Ref. 2 (see also Ref. 1) and dealt with ternary biquadratic forms.

the methods of the formal calculus of variations are called upon for the formation of differential invariants.

The study of the arithmetic theory of algebraic functions [14] made her more familiar with Dedekind's module and ideal theory, and this acquaintance would later give direction to her future work. In the joint work with Schmeidler [17] the module-theoretic concepts of direct sum and intersection decomposition, residue class modules, and module isomorphism, which run like red threads ("rote Fäden") through her later work, are developed and investigated. Here the uniqueness proofs are also given for the first time by means of the exchange method, and intersection representations are obtained by using a finiteness condition.

The first important consequence of this method was obtained in the now classic work of 1921, *Idealtheorie in Ringbereichen* [19]. In this, after the concepts of ring and ideal are defined, an equivalent finiteness condition, the ascending chain condition, is derived from the Hilbert theorem on the finite ideal basis. The representation of arbitrary ideals as intersections of primary ideals, which E. Lasker obtained for the case of polynomial rings by means of ideal theory, is recognized as following from the ascending chain condition alone. In addition to the concept of primary ideal (an abstract form of the Lasker concept, at the same time a generalization of the Dedekind concept of single-primed ideal), that of irreducible ideal is established and four uniqueness theorems are proved with the module method already mentioned.

This work forms the undeniable foundation of the present "general ideal theory." Its results needed to be extended in two different directions. First, it was necessary to obtain elimination theory as a corollary to the general ideal theory and to reestablish the theory of zeros of polynomial ideals from this point of view. In her handling of Hentzelt's elimination theory [22] and in two further works, Refs. 24 and 25, Emmy Noether struggled with this problem; but only in her lectures of 1923-1924 did she give the solution its final form. It speaks of her magnanimity that when I, in a followup to her work, one year later found the same foundation of the theory of zeros, she left its publication to me.

The second extension necessary was the establishment of the relevance of the general ideal theory to the classic Dedekind ideal theory of principal orders in number and function fields. It was necessary to find the conditions a ring must satisfy so that each of its ideals can be expressed not only as an intersection of primary ideals, but also as a product of prime ideal powers. This problem was also solved completely [31]. In addition to the finiteness conditions (ascending and descending chain conditions), the condition of "integral closure" appears in a fundamental way. With the aid of her older invariant-theoretic method she also arrived at a finiteness theorem for modular invariants by transferring the finiteness conditions to finite extensions of a ring.

The long ideal-theoretical papers, Refs. 19 and 31, form the basis of a long series of fruitful works, mostly by Emmy Noether's students, which W. Krull treated extensively in his monograph *Idealtheorie* (Ergebn. Math. *4*, 3, 1935).

Meanwhile, Emmy Noether herself was already working on a new set of problems. The same module concepts from which she developed commutative ideal theory would also show their power in the noncommutative. Above all it was possible to derive the representation theories of groups and hypercomplex systems [Ed. The modern terminology, from module theory, is "algebras."] Namely, to each representation of a system $\mathfrak{R}$ there corresponds, by means of linear transformations, a unique $\mathfrak{R}$-module, the *representation module*. Hence the equivalence concept in representation theory can be inferred from the module isomorphism concept; likewise the concepts of reducibility, irreducibility, and complete irreducibility prove to be module-theoretical concepts. The following theorem now crystallizes as the central theorem of representation theory: Each irreducible $\mathfrak{R}$-module is equivalent to an ideal of the ring $\mathfrak{R}$.

Emmy Noether had already developed this close relationship between representation theory, module theory, and ideal theory in 1924 in her lectures (cf. the notes Ref. 28); she also based her work on discriminants [32] on it. This relationship was first explained in complete clarity and generality, however, in the Göttingen lecture series of 1927-1928 and in the resulting paper [35]. This work also contains a systematic ideal theory of hypercomplex systems, culminating in the theorem: The semisimple hypercomplex systems in the sense of J. H. Maclagan-Wedderburn are direct sums of simple right ideals; their representations are likewise completely reducible. From these theorems the entire Frobenius representation theory was developed and even generalized. While the Frobenius representation theory originated from the field of complex numbers, the Noether theory allowed representations in arbitrary fields to be handled directly. The question of the relationships between the representations in different fields (the so-called arithmetical theory of groups of linear substitutions) then arose, particularly the question of *splitting fields*, in which a given representation decomposes into absolutely irreducible representations. In the Noether theory these questions are subsumed under the more general question of the structure of the product of two simple hypercomplex systems, which was solved completely by using module-theoretic methods [41]. Moreover, a characterization of the splitting field of a division algebra as a maximal commutative subfield of the algebra itself or of a complete matrix ring over this algebra was given in Ref. 33. This embedding of splitting fields gives at the same time a deep insight into the structure of the algebra itself: It can be represented as a "crossed product" of the splitting field with its Galois group.*

The simplest case of the crossed product is the "cyclic algebra," which results when the splitting field is cyclic and embedded in the algebra itself. The structure of such a cyclic algebra depends on whether certain elements of the ground field are norms of elements of the splitting field. In particular, if the ground field is an algebraic number field, then the norm theory of cyclic extensions is a topic in class field theory, which thus appears as being closely related to the theory of algebras [40]. The further utilization of this close relationship by Noether, H. Hasse, R. Brauer, and C. Chevalley in constant interaction led, on the one hand, to a new foundation for certain parts of class field theory with hypercomplex methods and, on the other hand, to the proof of a long-conjectured "principal theorem of the theory of algebras," which states that each division algebra over an algebraic number field is cyclic [39].

The consideration of arbitrary crossed products instead of cyclic algebras finally made possible the transfer of theorems of class field theory, in particular of the "principal genus theorem," to nonabelian fields [42].

One goal which Emmy Noether persistently pursued for many years, unperturbed by the skepticism of the number theorists, was achieved with the abstract elucidation of class field theory. The attainment of this goal, however, was by no means the end of her investigations. Indefatigable and in spite of all external difficulties, she proceeded along the way indicated by the concepts she formed. Also, when she lost her teaching rights in Göttingen in 1933 and was appointed to the women's college in Bryn Mawr (Pennsylvania), she soon gathered a school around her there and in nearby Princeton. Her research, which covered commutative algebra, commutative arithmetic, and noncommutative algebra, now turned to noncommutative arithmetic [43], but was suddenly terminated by her death.

As characteristic features we have found: An exceptionally energetic and consistent pursuit of abstract elucidation of the material to complete methodical clarity; a stubborn clinging to methods and concepts once they

---

*The Noether theory of crossed products is described by H. Hasse, Theory of cyclic algebras, *Trans. Amer. Math. Soc. 34*, pp. 180-200, as well as in the monograph of M. Deuring, *Algebren*, Ergebn. Math. *4*, 1, pp. 52-56.

had been acknowledged as being correct, even when they still appeared to her contemporaries as abstract and futile; an aspiration to classify all special relationships under specific general abstract models.

Indeed, her thoughts deviated in some respects from those of most other mathematicians. We are all so dependent on figures and formulas. For her these resources were useless, rather annoying. Her sole concern was with concepts, not with intuition or calculations. The German letters which she scribbled down hurriedly on the blackboard or on paper in typical simplified form were for her representations of concepts, not objects of a more or less mechanical calculation.

This totally unintuitive and unanalytical attitude was undoubtedly also one of the main causes of the complexity of her lectures. She had no didactical gifts, and the great pains she took to explain her remarks by quickly spoken interjections even before she had finished speaking were more likely to have the opposite effect. And still how exceptionally great was the impact of her talks, everything notwithstanding! The small, faithful audience, mostly consisting of a few advanced students and often just as many lecturers and foreign guests, had to exert themselves to the utmost to keep up. When that was done, however, one had learned far more than from the most excellent lecture. Completed theories were almost never presented, but usually those that were still in the making. Each of her lecture series was a paper. And nobody was happier than she herself when such a paper was completed by her students. Completely unegotistical and free of vanity, she never claimed anything for herself, but promoted the works of her students above all. For all of us she always wrote the introductions in which the main ideas of our work, which we initially never could understand and express in such clarity on our own, were explained. She was a faithful friend to us and at the same time a strict and unprejudiced judge. As such she was also invaluable to *Mathematische Annalen.*

As mentioned earlier, her abstract nonintuitive concepts initially found little acknowledgment. As the success of her methods also became clear to those of a different mind, this situation changed accordingly, and during the last eight years, prominent mathematicians from home and abroad went to Göttingen to ask her advice and listen to her lectures. In 1932 she shared the Ackermann-Teubner commemorative prize for arithmetic and algebra with E. Artin. And throughout the world today the triumphant progress of modern algebra which developed from her ideas seems to be unending.

5

In Memory of Emmy Noether

P. S. Alexandroff

Preface: On September 5, 1935, a ceremonial session of the Moscow Mathematical Society was held in honor of Emmy Noether, whose death had occurred on April 14. Besides the members of the Society, the meeting was attended by the brother of the deceased, Fritz Noether of Tomsk University, and the members of the First International Conference on Topology, then taking place in Moscow. After the members of the Society stood in honor of the deceased, the President, P. S. Alexandroff, delivered the address which is given below. Mathematical papers connected with Emmy Noether's work and ideas were then presented by John von Neumann (Princeton), Andre Weil (Paris), and A. G. Kurosh (Moscow). These papers are published in *Matematicheskii Sbornik.** In conclusion, Solomon Lefshetz (Princeton) spoke a few words in honor of the deceased, with particular emphasis on the great significance which the ideas of Emmy Noether had for the modern development of topology.

Translated by E. L. Lady, Department of Mathematics, University of Hawaii, Honolulu, Hawaii, with the permission of the editors of *Uspekhi Mathematicheskikh Nauk,* from *Uspekhi Mat. Nauk 2* (1936), 254-265.
*Editor's note: *Mat. Sbor. 436*, 5 (1936) consists of the proceedings of the First International Conference on Topology. It contains a paper by von Neumann and a paper and an abstract by Weil, but these do not appear to be related to Emmy Noether's work.

Emmy Noether died on the 14th of April of this year in the small Pennsylvania town Bryn Mawr (U.S.A.) at the age of 53, after undergoing surgery. Formerly a professor at Göttingen University, she was one of the major figures in the modern mathematical world.

The death of Emmy Noether is not only a loss to mathematics, but one which is in the full sense of the word tragic. The greatest woman mathematician of all time died at the height of her creative powers, after being driven out of her own country and torn away from the mathematical school she had created over a period of years, one of the most brilliant mathematical schools of Europe. She was also torn away from her family, which was scattered about the world because of the same political barbarity which had forced her to leave Germany.

The Moscow Mathematical Society today sadly pays its respects to one of its most distinguished associates, who for more than ten years maintained close ties of constant scientific cooperation, sincere sympathy, and heartfelt friendship with the Society, with the Moscow mathematical world, and with the mathematicians of the Soviet Union. Fritz Noether, brother of the deceased, formerly professor at the Brevslavl Technical School and now professor at the Tomsk Mathematical Institute, is here with us today. Allow me to express my deep condolences to him.

The biography of Emmy Noether is not very complicated. Born in Erlangen on March 23, 1882, she was the daughter of the noted mathematician Max Noether. Her mathematical talent developed slowly. Her dissertation, written in Erlangen under Gordan and completed in 1907, was in the spirit of Gordan's formally computable invariant theory. She frequently recalled this dissertation thereafter, always referring to it with deprecatory terms such as "Formelngestrüpp" and "Rechnerei." For all this, one should note that Emmy Noether, that fervent enemy of anything computational or algorithmic in mathematics, was fully capable of mastering that method. This is shown not only by her dissertation (which was not in fact a major contribution to mathematics) but by her subsequent work on differential invariants (1918) which thereafter became classic. But already in these works the fundamental character of her mathematical talent was becoming manifest: the striving toward an abstract formulation of mathematical problems and the ability to find exactly that formulation which brings to light the true logical nature of the given question, freed from all those accidental details which complicate and obscure the true situation.

Her work on differential invariants was done in Göttingen, where she came in 1916. Her research at this time was strongly influenced by Hilbert. It is often forgotten that during this period Emmy Noether obtained first-rate results on Hilbert's type of concrete algebraic problems. These results and her work on differential invariants would in themselves have been enough to establish her reputation as a first-class mathematician and were a contribution to mathematics hardly inferior to the well-known work of Sonia Kowalevski. But when Noether's mathematics is discussed, one has in mind not these early works, however, significant their specific results were, but the whole fundamental period of her mathematics beginning in about 1920. She then appeared as the creator of a new direction in algebra and became the leader, the most consistent and brilliant representative, of a particular mathematical doctrine—of all that is characterized by the term *begriffliche Mathematik*.

Emmy Noether was herself partly responsible for her early work being remembered less frequently than would be natural. For with all the fervor of her nature, she was herself ready to forget what had been done in the first years of her mathematical activity, considering these results as standing apart from her true mathematical path—the creation of a general abstract algebra.

It is not my place here to elucidate everything done in mathematics by Emmy Noether. First of all, not being an algebraist, I don't feel qualified for such a task. Secondly, within the limitations of a eulogy, this has been done brilliantly and fully competently by Hermann Weyl at the Ceremonial Session in honor of Emmy Noether on April 26, 1935 at Bryn Mawr [published in *Scripta Mathematica* 3(3), June 1935] and also by van der Waerden (*Mathematische Annalen 111*, 1935, p. 469). My task today is different. I simply wish to portray the deceased for you as vividly as possible, as a mathematician, as the leader of a great mathematical school, and as a brilliant, original, and charming personality.

Emmy Noether's own distinctive approach to mathematics began to be seen in 1919-1920. She herself considered her joint work with W. Schmeidler (*Mathematische Zeitschrift 8*, 1920) as marking the beginning of this basic period of her activity. That work served as a sort of prologue to her general theory of ideals, developed in 1921 in her classical memoir *Idealtheorie in Ringbereichen*. I think that, of all Emmy Noether's work, it is the foundation of the general theory of ideals, and everything connected with it, which exerted, and will continue to exert, the greatest influence on mathematics as a whole. These ideas have not only already had a number of concrete applications—for instance in van der Waerden's work on algebraic geometry— but they, above all, exerted an essential influence on algebra itself, and in some respects also on the mathematical thought of our epoch as a whole. If it is true beyond any doubt that contemporary mathematics marches under the banner of algebraization, the penetration of algebraic concepts and methods into the most diverse mathematical theories, then this became possible only after Noether's work. It was she who taught us to think in terms of elementary, and therefore, general, algebraic concepts—homomorphisms, groups and rings with operators, ideals—rather than in terms of computational algebra. And she therefore led us to discover unifying algebraic principles in places where they had previously been obscured by complex special circumstances which the classical mathematician did not recognize as algebraic. She was the first to formulate the basic homomorphism and isomorphism theorems, the ascending and descending chain conditions, and the concept of a group with operators. These have now become powerful tools, in constant daily use in a multitude of mathematical fields completely unrelated in subject matter to the work of Noether herself. One need only consider Pontryagin's work on topological groups, Kolmogorov's recent work on the combinatorial topology of locally compact spaces, and the work of Hopf on the theory of continuous mappings, not to mention van der Waerden's work on algebraic geometry, to feel the influence of her ideas. One can see it even in Hermann Weyl's book *Gruppentheorie und Quantenmechanik.*

I have been referring intentionally to mathematical areas not directly investigated by Noether. As to algebra itself, including group theory, here she created an entire new trend, with a number of very talented mathematicians continuing her work along completely concrete lines. In particular, among the more notable attainments growing from the soil of Noether's general ideal theory are the general formulation of elimination theory and the theory of algebraic manifolds. In connection with this, I would like to mention especially that among the most significant algebraic work during the past decade here in Moscow have been Schmidt's well-known results on the uniqueness of direct sum decompositions for groups, as well as a number of works by Kurosh very much influenced by Noether.

For all the concreteness and constructiveness of some of Emmy Noether's work in various periods of her creative life, there is no doubt that her fundamental effort and passion was directed toward generalization, and especially toward axiomatic concepts of great generality. Now is an appropriate time to consider this side of her work more carefully, because, among other things, the question of the general versus the specific, the abstract versus the concrete, the axiomatic versus the constructive, is currently one of the most burning issues in the practice of mathematics. We see that, on the one hand, mathematical journals are undoubtedly burdened needlessly with a mass of every possible sort of generalizing, axiomatizing papers, often devoid of any concrete mathematical content. And on the other hand, here and there one sees proclamations that the only *real* mathematics is "classical." Under this battle cry, we see important mathematical problems swept aside only because they are contrary to some mental habit or other, or because they involve concepts which were unknown a few decades ago, such as that of a ring or field, of a topological or function space, to name only a few. Hermann Weyl addresses this general question in the eulogy I have already cited. What he says on this score penetrates so completely to the heart of the matter that I cannot refrain from quoting him in full:

> In a conference on topology and abstract algebra as two ways of mathematical understanding, in 1931, I said this:
> "Nevertheless I should not pass over in silence the fact that today the feeling among mathematicians is beginning to spread that the fertility of these abstracting methods is approaching exhaustion. The case is this: that all these nice general notions do not fall into our laps by themselves. But definite concrete problems were first conquered in their undivided complexity, single-handedly by brute force, so to speak. Only afterwards the axiomatician came along and stated: Instead of breaking in the door with all your might and bruising your hand, you should have constructed such and such a key of skill, and by it you would have been able to open the door quite smoothly. But they can construct the key only because they are able, after the breaking in was successful, to study the lock from within and without. Before you can generalize, formalize, and axiomatize, there must be a mathematical substance. I think that the mathematical substance in the formalizing of which we have trained ourselves during the last decades becomes gradually exhausted. And so I foresee that the generation now rising will have a hard time in mathematics."
> Emmy Noether protested against that: and indeed she could point to the fact that just during the last years the axiomatic method had disclosed in her hands new, concrete, profound problems . . . and had shown the way to their solution.

In this, there is much worth noting. First of all, one cannot argue with the point that every axiomatic treatment of mathematical material should be preceded by a concrete and, I would say, naive understanding of it, and that, furthermore, axiomatization is interesting only when it deals with real mathematical knowledge (the "mathematical substance" mentioned by Weyl) rather than, to put it bluntly, fancy wrapping paper around empty boxes. This is clear, and certainly Emmy Noether would have had no protest. But she would definitely have protested against the pessimism that shows through the last words that Weyl quotes from his own 1931 speech. Emmy Noether was firmly convinced that the substance of human knowledge, mathematical and otherwise, is inexhaustible, at least in any reasonable time period. The "substance of recent decades" is being exhausted, but not mathematical substance as a whole, which has thousands of ties to everyday life and reality. This connection of all *great mathematics*, even the most abstract, with real life, was something that Noether felt intensely, even if she did not formulate it philosophically, with all her nature as a great scientist and a person who was alive and not embalmed in some abstract scheme. For her, mathematics was always knowledge of the world, never a game with meaningless symbols, and she warmly protested when real knowledge was claimed for only representatives of those mathematical fields directly related to applications. In mathematics, as in knowledge of reality, both aspects are equally valuable: the accumulation of individual facts with unique concrete constructions and the establishment of general principles which go beyond the uniqueness of each fact and translate knowledge of details into a new stage of axiomatic knowledge.

A deep feeling of reality underlay all Emmy Noether's work. Her whole mathematical personality was set against the attempt found in many mathematical circles to turn mathematics into some form of witty and original sport. In my many conversations with her on the nature of mathematical knowledge and creativity (conversations carried on for the most part on a naive level rather than in genuine philosophic terms) she more than once expressed sympathy with my favorite citation from Laplace: Si l'homme s'était borné à recueillir des faits, la science ne serait qu'une nomenclature stérile, et jamais il n'eut connu les grandes lois de la nature. In these words, spoken by one of the most concrete representatives of the exact sciences, a scientist with both feet on the ground of reality, we see the whole program of interaction of the concrete and the abstract in all of human knowledge, including mathematics. And it seems to me that Noether's work has fulfilled that program.

In 1924-1925, the school of Emmy Noether made one of its most brilliant acquisitions—the Amsterdam student van der Waerden, then completing his dissertation. He was 22 years old and one of the most gifted young mathematicians in Europe. Van der Waerden quickly mastered Noether's theories, augmenting them with essential new results of his own, and he more than anyone else helped spread her ideas. In 1927 he taught a course at Göttingen on the general theory of ideals and it had enormous success. The triumph of Noether's ideas, expressed so brilliantly by van der Waerden, spread from Göttingen to other leading European mathematical centers. It was no accident that Noether required a popularizer for her ideas: her lectures were designed for a small circle of students working in the same direction as she and who were in constant attendance. They were by no means suitable for a large mathematical audience. Noether was a poor lecturer, hurried and seemingly confused. But the power of her mathematical thought was great, and she possessed an unusual enthusiasm and passion. The same was true of her addresses at meetings and conferences. To the mathematician who was interested in her work and had already mastered her ideas, these addresses were very valuable. But those with more casual interest often could understand her exposition only with great effort.

Since 1927 the influence of Noether's ideas on modern mathematics has been constantly increasing, along with her own fame as a scientist. The direction of her work had been changing during this time, moving more and more into the realm of noncommutative algebra, representation theory, and the arithmetic theory of algebras. We see this in the two most fundamental works of the last period of her life, *Hyperkomplexe Grössen und Darstellungstheorie* (1929) and *Nichtkommutative Algebra* (1933), both published in *Mathematische Zeitschrift* (volumes 30 and 37). These works and those related to them immediately provoked a great response from algebraic number theorists, especially from Hasse. Among her students in this period, the most prominent was M. Deuring, but there were also a host of young beginning mathematicians, such as Witt, Fitting, and others.

Deuring gave an overview of Noether's work on algebras in his monograph *Algebren*, published in the *Ergebnisse der Mathematik* series.

Emmy Noether lived to see complete recognition for her ideas. In 1923-1925 she had to demonstrate the importance of the theories she was developing, but at the International Congress of Mathematicians in Zurich in 1932 she was crowned with laurels. The long expository lecture she gave there was a real triumph for her mathematics, and she could look back on the road she had traveled not only with inner satisfaction, but with an awareness of complete and unconditional recognition for her efforts. The Zurich Congress was the high point of her international mathematical career. Within a few months, German culture, and in particular Göttingen University, where it had been nurtured for centuries, would see the outbreak of Fascism, which within a few weeks would scatter to the winds everything that had been created over so many decades. One of the greatest tragedies experienced by human culture since the time of the Renaissance was taking place, a tragedy which a few years before would have seemed an impossibility in twentieth century Europe. One of its manifold victims was the Noether algebra school in Göttingen. Noether was driven from the university walls. Deprived of the right to teach, she was forced to emigrate. She accepted the invitation of the women's college at Bryn Mawr (1933) and spent the last year and a half of her life there.

Emmy Noether's career was paradoxical and will forever stand as an example of the disgraceful stagnation and prejudice of the Prussian academic world and bureaucracy. She was given the title Privatdozent only at the insistence of Hilbert and Klein, after overcoming extraordinary resistance from the reactionary university circles. The basic official objection was the candidate's sex. "How can a woman be allowed to be a Privatdozent? Once she was a Privatdozent, after all, she could become a professor and member of the University Senate. Can we allow a woman into the Senate?" This protest provoked Hilbert's well-known reply: "Meine Herren, der Senat is ja keine Badenanstalt, warum darf eine Frau nicht darin?" In fact, the opposition of influential members of the unversity's reactionary circles was provoked not so much by the fact that Noether was a woman as by her well-known extreme radical political convictions. The situation was aggravated, from their point of view, by the fact that she was a Jew. But I will have more to say about that later.

She did finally obtain the title Privatdozent, and later on that of Adjunct Professor. As a result of Courant's efforts, she was given a so-called Lehrauftrag, i.e., a small honorarium (200-400 marks per month) for the courses she taught. This had to be approved every year by the Ministry. Such was her situation, without even a guaranteed salary, up to the moment when she was driven out of the university and forced to emigrate. She was not a member of a single academy, not even the academy of the city where most of her work was carried out. Here is what Hermann Weyl says on that score in his eulogy:

> When I was called permanently to Göttingen in 1930,[1] I earnestly tried to obtain from the Ministerium a better position for her, because I was ashamed to occupy such a preferred position beside her whom I knew to be my superior as a mathematician in many respects. I did not succeed, nor did an attempt to push through her election as a member of the Göttinger Gesellschaft der Wissenschaften.[2] Tradition, prejudice, external considerations, weighted the balance against her scientific merits and scientific greatness, by that time denied by no one. In my Göttingen years, 1930-1933, she was without doubt the strongest center of mathematical activity there, considering both the fertility of her scientific research program and her influence upon a large circle of pupils.

It would be hard for me to add anything to Weyl's words.

Emmy Noether had close ties with Moscow. These ties began in 1923 when the late Pavel Samuelovich Urysohn and I first arrived in Göttingen and immediately fell in with the mathematical circle of which Noether was the head. We were immediately struck with the fundamental traits of the Noether school: the mathematical enthusiasm of its leader, which she conveyed to all her students, her deep confidence in the importance and the mathematical productiveness of her ideas—a confidence which was far from universally shared even in Göttingen—and the unusual straightforwardness and sincerity of the relations between Noether and her students. At the time, this school consisted almost entirely of young students from Göttingen. The period when it would become international in its composition and be recognized as the primary center of algebraic thought in the world was still in the future.

The mathematical interests of Emmy Noether, who was then in the very midst of her work on the general theory of ideals, and the interests of myself and Urysohn, then centering around so-called abstract topology, had many points of contact and soon led to continual, almost daily, mathematical discussions. Noether, however, was interested not only in our topological work, but in everything going on in the Soviet Union, both in mathematics and otherwise. She did not conceal her sympathies toward our country and its governmental and social organization, despite the fact that the manifestation of these sympathies seemed both outrageous and in poor taste to most of those in European academic circles. It came to the point where Noether was literally driven out of one of the Göttingen pensions (where she lived and ate) by the demands of student leaders in the pension who did not want to live under the same roof with "a Marxist-leaning Jewess." That was an apt prologue to the drama which would end her life.

Emmy Noether was sincerely delighted by Soviet scientific, and especially mathematical, successes, because she saw in them the refutation of all the predictions that "the Bolsheviks are destroying culture." She felt the dawn of a great new culture already in the works. She, a representative of one of the most abstract areas of mathematics, had an astonishing sensitivity to the great historical exploits of our era. She always had a lively interest in politics and hated war and chauvinism in all its forms with her whole being. Her sympathies were always unwaveringly with the Soviet Union, in which she saw the beginning of a new era in history and a firm support for everything progressive. This trait was such a bright one of Noether's character and it cast such an imprint on her whole personality, that to be silent about it would be to tendentiously distort her image as a scholar and as a human being.

My mathematical and personal friendship with Emmy Noether began in 1923 and did not cease up to the date of her death. In his eulogy, Weyl referred to this as follows: "She held a rather close friendship with Alexandroff in Moscow. I believe that her mode of thinking has not been without influence upon Alexandroff's investigations." I am glad to be able now to confirm Weyl's supposition. Noether's influence both on my own research and on that of other Moscow topologists was great and touched on the very essence of our work. In particular, my theory of continuous decompositions of topological spaces was created in large part under the influence of my conversations with her in December and January of 1925-1926 when we were in Holland together. It was also at this time that Noether's first ideas on the set-theoretic foundations of group theory were developing. She lectured on these in the summer of 1926. Although she returned to them several times later, these ideas were not developed further in their initial form, probably because of the difficulty of axiomatizing the concept of a group by taking the coset decomposition as the basic notion. But the idea of set-theoretic analysis of the concept of a group turned out to be fruitful, as shown by the subsequent work by Ore, Kurosh, and others.

Later years brought a strengthening and deepening of Noether's interest in topology. In the summers of 1926 and 1927 she attended courses given by myself and Hopf in Göttingen. She quickly found her way in what was for her a completely new field and made constant comments, some of which were deep and subtle. On first learning the systematic construction of combinatorial topology from our lectures, she immediately observed that it would be useful to consider directly the groups of chains and cycles of a given polyhedron. She proposed that instead of the usual definition of Betti numbers and torsion coefficients one should immediately define the Betti *group* as the factor group of the group of cycles modulo the subgroup of bounding cycles. This observation now seems self-evident. But in those years (1925-1928) this was a completely new point of view whose reception by many eminent topologists was far from sympathetic. Hopf and I immediately saw things Noether's way, but for some time in this respect we were a minority. Of course now it would scarcely occur to anyone to construct combinatorial topology without using abelian group theory as the foundation—all the more reason to note that the idea was originally Noether's. She immediately observed how simple and transparent the proof of the Euler-Poincaré formula became using the Betti group concept systematically. These remarks inspired Hopf to completely reprove the well-known fixed point theorem which Lefschetz had proved for manifolds and Hopf had generalized to arbitrary polyhedra. Hopf's paper *Eine Verallgemeinerung der Euler-Poincaréschen Formel*, published in the *Göttinger Nachrichten* in 1928, bears the stamp of Noether's remarks.

Noether spent the winter of 1928-1929 in Moscow. She gave a course in abstract algebra at Moscow University and conducted a seminar in algebraic geometry at the Communist Academy. She quickly made contacts with most Moscow mathematicians, in particular with L. S. Pontryagin and O. Ju. Schmidt. The strong algebraic note in Pontryagin's work undoubtedly gained a great deal in its development from his contacts with Noether. She quickly fell in with our Moscow ways, both in mathematics and in everyday life. She lived in a modest room in the KSU dormitory at the Crimean Bridge and generally walked to the University. She was very interested in our way of life, and in particular in the life of Soviet youth, especially the students.

In the winter of 1928-1929, as usual, I traveled back and forth to Smolensk, and I gave lectures on algebra at the Pedagogical Institute there. Inspired by my constant conversations with Noether, I oriented my lectures that year around her concepts. Among my audience there, A. G. Kurosh immediately singled himself out, finding the Noetherian theory I was expounding much to his taste. And it was thus through my teaching that Emmy Noether acquired a student who has since—as everyone knows—grown to the status of an independent mathematician, whose work since that time has been carried on in the circle of ideas created by her.

In the spring of 1929 she left Moscow for Göttingen with the firm intention of returning the following summer. Several times she was close to the accomplishment of this intention, closest of all in the last year of her life. After being driven out of Germany, she seriously considered coming to Moscow for good, and I corresponded with her on that topic. She understood perfectly that nowhere else would she find the same opportunities to create a splendid new mathematical school that would replace the one that had been taken away from her in Göttingen. And I was already negotiating with the Commissariat of Education about obtaining a chair for her at Moscow University. However, as usual, the Commissariat was slow to make its decision and did not give me a final answer. Meanwhile, time was passing, and Emmy Noether, deprived of even that modest stipend which she had formerly received, could not wait and had to accept the invitation of the women's college at Bryn Mawr.

The death of Emmy Noether has deprived us of one of the most charming human beings that I have ever had the pleasure to encounter. Her kindness was unusual. Any sort of posing or insincerity was foreign to her. Her simplicity and joy in living and her capacity to ignore everything in life which is not essential created that atmosphere of warmth, calm, and good feeling around her that will never be forgotten by anyone who ever had anything to do with her. And yet her kindness and softness never crossed the line to become mushiness and mere agreeability. She had her own opinions and knew how to defend them vigorously and insistently. Her nature, for all its mildness, was passionate, temperamental, and willful, and she always spoke her mind frankly without worrying about people's objections. Her feelings for her students were touching. They provided the basic environment in which her life took place, taking the place of her absent family. Her concern for all their needs, both mathematical and otherwise, her sensitivity and responsiveness were exceptional. Her great sense of humor, which made social gatherings and personal contacts with her so pleasant, enabled her to counter the injustices and absurdities that beset her academic career easily and without anger. In such circumstances, instead of being offended she would simply laugh. But she was very offended indeed, and protested sharply, when even the smallest injustice was directed at one of her students. All her maternal feelings were bestowed on them.

Social, kind, and direct in her dealings with people, she knew how to join effusiveness with calm and the absence of any fuss. Vanity and ambition for external success were foreign to her. But for all that, she knew her value and fiercely insisted on her right to scholarly influence.

In her home, which is to say in the garret where she lived in Göttingen (Frieländerweg 57), she was an enthusiastic party giver. People of every scientific stature, from Hilbert, Landau, Brauer, and Weyl to quite young students, met there and felt at ease, which was hardly the case in many other European scientific salons. These "idle evenings" were arranged for the most diverse reasons—in the summer of 1927, for instance, because of the frequent visits of her student van der Waerden from Holland. Noether's parties, and likewise her excursions into the country, were a bright and unforgettable feature of the Göttingen mathematical life of the entire decade from 1923 to 1932. Many lively mathematical discussions were carried on at these parties, but there was also a lot of just good fun and joking, and sometimes good Rheinwine and other delicacies.

Such was Emmy Noether, the greatest of women mathematicians, a great scholar, an astonishing teacher, and an unforgettable person. In her there was nothing of the "lady scientist" in quotation marks, the "blue stocking." True enough, Weyl said in his euology that "the Graces hardly stood by her cradle," and that is correct, if one is thinking of the well-known heaviness of her figure. But in the very same lines Weyl speaks of her not only as a great mathematician but as a magnificent woman. And she was that. Her femininity was manifest in that mild and subtle lyricism which was the basis for her wide-ranging but never superficial interests, in people, in her own specialty, and in the concerns of all humankind. She loved people, science, and life with all the fervor, joyousness, unselfishness, and affection of which a deeply feeling—and feminine—soul was capable.

NOTES

1. In 1930 H. Weyl received Hilbert's chair at Göttingen University after Hilbert retired upon reaching the age limit (68). Hilbert's chair was considered the foremost mathematical chair in Germany and carried with it an augmented salary. When Weyl emigrated to America in 1933 it was given to Hasse.
2. The Göttingen Academy of Sciences, founded in 1742, in Hanoverian times, was called, following the model of the British academies, the Royal Society of Sciences—the Königliche Gesellschaft der Wissenschaften zu Göttingen.

III

NOETHER'S MATHEMATICS

6

Galois Theory

Richard G. Swan

1. INTRODUCTION

Noether's contributions to Galois theory center around the following problem:

Question A.   Let F be any number field and G any finite group. Is there a Galois extension K of F with Galois group isomorphic to G?

The answer has long been known to be affirmative for symmetric groups. Šăfarevič (1954) showed that it is true for all solvable groups but the general problem is still unsolved. Fried and Kollar (1978) showed that G can be the automorphism group of a nonnormal extension of F. A completely new and very promising approach to the problem has recently been discovered by D. Harbater.

Although Noether's attempt to solve this problem was unsuccessful, it has led to a number of very fruitful developments which have revealed unexpected connections with problems in class field theory, algebraic geometry, and topology.

Let $A = k[x_1, \ldots, x_n]$ be a polynomial ring over a field k and let G be the symmetric group on n letters permuting the x_i. The classical theorem on symmetric functions asserts that the fixed subring A^G is a polynomial ring on the elementary symmetric functions $\sigma_1, \ldots, \sigma_n$. Hilbert (1892) showed that if k is a number field, one could specialize the σ_i to elements $a_i \in k$ in such a way that A specializes to a Galois extension $K = A \otimes_{A^G} k$ with group G.

Noether's idea was to extend this to other groups G by letting G act as a permutation group on $x_1, \ldots, x_n$. Here it is not hard to give examples where A^G is not a polynomial ring. However, Noether observed that it would suffice to know that the quotient field of A^G is a rational (i.e., purely transcendental) extension of k. This led to the following well-known problem often referred to as Noether's conjecture, although as Lenstra (1974) points out, she only stated it as a question.

Question B. Let G be a permutation group acting on indeterminates $x_1, \ldots, x_n$. Is $k(x_1, \ldots, x_n)^G$ a rational extension of k?

An affirmative answer for a given G and number field k would imply that G can be realized as a Galois group over k. Unfortunately, as we shall see, the answer is negative even for a cyclic group.

Remark. Noether's name is sometimes also associated with the generalization of Hilbert's Theorem 90 to noncyclic extensions. This appears in Artin's book (1944) under the name "Noether's equations." I do not know the origin of this terminology. Noether herself [42]* attributes the result to Speiser (1919).

2. CONSTRUCTING GALOIS EXTENSIONS

Let G be a finite group acting faithfully on a finite set of indeterminates $x_1, \ldots, x_n$. Let k be a number field, $K = k(x_1, \ldots, x_n)$, and $E = K^G$.

Theorem 6.1 [6] If E is a rational extension of k then G can be realized as the group of a Galois extension of k.

Proof. Let $E = k(y_1, \ldots, y_n)$ with the y_i algebraically independent. Let $K = E(\alpha)$ and let $f(y_1, \ldots, y_n, z) = 0$ be the minimal equation for α over E. By Hilbert's irreducibility theorem (Hilbert 1892, Lang 1962) we can find $a_1, \ldots, a_n \in k$ so that $f(a_1, \ldots, a_n, z)$ is defined and irreducible over k. The splitting field of this gives the required extension. This can be seen by using the algorithm for computing Galois groups given by van der Waerden (1953).

An alternative approach which gives a very clear picture of the situation has recently been developed by D. Saltman (1980). This uses the Galois

---

*Numbered references are to Noether's papers listed at the end of this book.

theory of rings. Let G be a finite group acting by automorphisms on a commutative ring B. Let $A = B^G$. If A and B are fields, B will be a Galois extension of A with group G provided that G acts faithfully on B. In the general case, we need a stronger condition, namely, that G acts faithfully on B/I for all proper G-invariant ideals I. This is equivalent to the "principal homogeneous space" condition $B \otimes_A B \overset{\approx}{\to} \Pi_{\sigma \in G} B$ by $b_1 \otimes b_2 \mapsto (b_1 \sigma(b_2))$, and also to the condition that there exist $u_i, v_i \in B$ with $\Sigma u_i \sigma(v_i) = \delta_{\sigma 1}$ where $\delta_{11} = 0$, $\delta_{\sigma 1} = 0$ for $\sigma \neq 1$ (Chase, Harrison, and Rosenberg 1969; De Meyer and Ingraham 1971).

In the situation considered above we can find rings A and B finitely generated over k with quotient fields E and K and such that B is a Galois extension of A with group G. We start with $B = k[x_1, \ldots, x_n]$, $A = B^G$, and replace A and B by $A[s^{-1}]$, $B[s^{-1}]$ so that the elements u_i, v_i for K/E lie in $B[s^{-1}]$. By a further localization if necessary, we may assume that $A = k[y_1, \ldots, y_n, g^{-1}]$ with $g \in k[y_1, \ldots, y_n]$.

Now if $a_1, \ldots, a_n \in k$ and $g(a_1, \ldots, a_n) \neq 0$, define $A \to k$ by $y_i \mapsto a_i$. Then $k \otimes_A B$ will be a Galois extension of k. However, it may not be a field but only a product of fields. To avoid this, consider α and $f(y_1, \ldots, y_n, z)$ as defined above. By localizing A and B again we can assume that $\alpha \in B$ and $f \in A[z]$. If we choose the a_i so that $f(a_1, \ldots, a_n, z)$ is irreducible over k, then $k \otimes_A B$ contains the field $k(\beta)$ where $\beta = 1 \otimes \alpha$ is a root of $f(a_1, \ldots, a_n, z)$. Comparison of degrees then gives $k \otimes_A B = k(\beta)$.

The rationality of E over k is needed for the application of Hilbert's irreducibility theorem since it is not clear that we can specialize the y_i to the a_i unless the y_i are algebraically independent.

3. GENERIC GALOIS EXTENSIONS

In Ref. 11 Noether showed that the ideas of Section 2 above could be used to parameterize all Galois extensions of k with group G when Question B has an affirmative answer. Her result excluded certain "singular" cases. This restriction was later removed by Kuyk (1964). Saltman (1980) has given a very elegant formulation of Noether's idea.

Definition (Saltman 1980). Let k be a field and G a finite group. A generic Galois extension for k with group G is a Galois ring extension $A \subset B$ of commutative k-algebras with group G such that $A = k[y_1, \ldots, y_n, g^{-1}]$ is a localization of a polynomial ring and such that the following property holds:

If F is a field containing k and L/F is a Galois field extension with group G, there exists a k-algebra homomorphism $A \to F$ so that $L \cong F \otimes_k B$ as a Galois extension.

Thus any Galois field extension of F with group G can be parameterized by the images $a_1, \ldots, a_n$ of $y_1, \ldots, y_n$ in F.

Theorem 6.2 Let k be an infinite field. Let G act faithfully on a set of indeterminates $x_1, \ldots, x_n$. If $k(x_1, \ldots, x_n)^G$ is a rational extension of k then there is a generic Galois extension for k and G.

Proof: We choose $A \subset B$ as in Section 2. If we can find a G-equivariant map $\varphi : B \to L$ then φ sends A to F and we get $F \otimes_A B \to L$. If, in addition, G acts faithfully on $\{\varphi(x_i)\}$, then $F \otimes_A B \to L$ will be onto since L/F is Galois and hence $F \otimes_A B \xrightarrow{\cong} L$ by comparing degrees. The following result of Kuyk gives the required φ.

Lemma 6.3 (Kuyk 1964) Let F be an infinite field and let L/F be a Galois field extension with group G. Let G act on a finite set $X = \{x_1, \ldots, x_n\}$ and let $h(x_1, \ldots, x_n) \in F[x_1, \ldots, x_n]$ be a nonzero polynomial. Then there is an equivariant embedding $\varphi : X \to L$ such that $h(\varphi(x_1), \ldots, \varphi(x_n)) \neq 0$.

Proof: If H_i is the stabilizer of x_i we send x_i to some $\alpha_i \in L^{H_i}$ and extend equivariantly. Since F is infinite we can make $h(\varphi(x_1), \ldots, \varphi(x_n)) \neq 0$ provided $\varphi(\alpha_1, \ldots, \alpha_n)$ is a nonzero polynomial in the coordinates of the α_i. It is sufficient to check this after extending the ground field from F to L and it is then easy since $L \otimes_F L \cong L^n$. To make φ injective, replace h by $h \cdot \Pi(x_i - x_j)$.

Unfortunately, generic Galois extensions do not always exist. Saltman (1980) shows that there is no generic Galois extension over $\mathbf{Q}$ for a cyclic group G of order $8n$. He observes that if we could take the unramified extension of $\mathbf{Q}_2$ of degree $8n$ and parameterize it as above, then by approximating the parameters by rational numbers we could find a cyclic extension K of $\mathbf{Q}$ of degree $8n$ such that 2 remains prime in K. This is impossible by Wang's well-known counterexample to Grunwald's theorem in class field theory (Wang 1948).

In particular, for such G, $\mathbf{Q}(x_1, \ldots, x_{8n})^G$ cannot be rational over $\mathbf{Q}$, where G permutes the x_i transitively.

4. FISCHER'S THEOREM

Soon after Noether formulated Question B, the following partial answer was given by E. Fischer (1915, 1916).

Galois Theory

Theorem 6.4 Let B be a finite abelian group of exponent e acting on the finite set $\{x_1, \ldots, x_n\}$. Let k be a field of characteristic prime to e and containing all eth roots of unity. Then $k(x_1, \ldots, x_n)^G$ is rational over k.

Proof: Let $V = \Sigma k x_i$. Then V is a sum of 1-dimensional G-invariant subspaces $V = \oplus V_j$ where $V_j = k y_j$, $\sigma y_j = \chi_j(\sigma) y_j$. Let $Y \subset k(x_1, \ldots, x_n)^*$ be the multiplicative group generated by the y_j. Let $\hat{G}$ be the character group of G and define $\varphi: Y \to \hat{G}$ by $\varphi(y_j) = \chi_j$. Then $M = \ker \varphi$ is free abelian of rank n and it is easy to see that $k(x_1, \ldots, x_n)^G = k(z_1, \ldots, z_n)$ where $\{z_i\}$ is a base for M, since $k(x_1, \ldots, x_n)^G$ is the quotient field $k(M)$ of the group ring $k[M]$.

5. GALOIS DESCENT

Masuda (1968) had the idea of extending k to make Fischer's theorem apply and using the result to obtain information about $k(x_1, \ldots, x_n)^G$.

Suppose G is an abelian group of exponent e prime to char k. Let $k' = k(\zeta_e)$ where ζ_e is a primitive eth root of 1. Let π be the Galois group of k'/k. By Fischer's theorem we have $k'(x_1, \ldots, x_n)^G = k'(M)$ where M is obtained from $V = \oplus k' x_i$ as in Section 4. Now V is clearly stable under π. Write $V = \oplus V^{(\chi)}$ where G acts on $V^{(\chi)}$ through the character χ. Then π permutes the $V^{(\chi)}$. If π_χ is the stabilizer in π of $V^{(\chi)}$, then $V^{(\chi)}$ has a base fixed by π_χ. This follows from a well-known result of Speiser (1919).

Lemma 6.5 Let k'/k be a Galois extension with group π. Let V be a vector space over k' with π-action such that $\sigma(rv) = \sigma(r)\sigma(v)$. Then $k' \otimes_k V^\pi \to V$ is an isomorphism.

To prove this, extend k to make k' split. This reduces the lemma to the trivial case $k' = k \times \cdots \times k$.

By choosing a base for $V^{(\chi)}$ fixed under π_χ we can produce a base y_i for V as in Section 4 such that π permutes the y_j. Now $0 \to M \to Y \to \hat{G}$ is a sequence of π-modules and we see that $k(x_1, \ldots, x_n)^G = (k'(x_1, \ldots, x_n)^G)^\pi$ can be obtained by the following construction:

Let k'/k be a Galois field extension with Galois group π. Let M be a π-module which is free and finitely generated as an abelian group. Define $Q(k'/k, M) = k'(M)^\pi$, the quotient field of $k'[M]^\pi$.

Note that the above lemma shows that $k' \otimes_k Q(k'/k, M) \stackrel{\sim}{\Rightarrow} k'(M)$.

It also follows easily from Lemma 6.5 that $Q(k'/k, M)$ is rational over k if M is a permutation module (Lenstra 1974, Kervaire 1973-1974).

6. A BIRATIONAL INVARIANT

Masuda (1968) used the approach of Section 5 to give an affirmative answer to Question B in certain cases. In 1969 I gave a partial converse to Masuda's result, which led to the fact that $Q(x_1, \ldots, x_{47})^G$ is not rational for $G = \mathbf{Z}/47\mathbf{Z}$. I will discuss here a sharper form of the invariant used in my paper (Swan 1969), which was introduced by Voskresenskii (1970).

Let K/k be a finitely generated field extension with k algebraically closed in K. Let k'/k be a finite Galois extension with group π. Let $K' = k' \otimes_k K$. Suppose we can find a finitely generated k'-subalgebra $A \subset K'$ with quotient field K' such that A is a unique factorization domain stable under π. Then we can use the π-module A^*/k'^* to produce a birational invariant for K. If B is another such ring we can find $a \in A^\pi$, $b \in B^\pi$ with $A[a^{-1}] = B[b^{-1}]$. Therefore, it will suffice to compare the modules resulting from A and $A[a^{-1}]$. Let $(p_1), \ldots, (p_n)$ be the prime factors of a in the UFD A. These are permuted by π. If S is the module of formal $\mathbf{Z}$-linear combinations of $(p_1), \ldots, (p_n)$, we have $0 \to A^* \to A[a^{-1}]^* \xrightarrow{j} S \to 0$ where $j(x) = \Sigma \text{ ord}_{p_i}(x)(p_i)$. This suggests considering the equivalence relation generated by $M \sim N$ if $0 \to M \to N \to S \to 0$ where S is a permutation module, i.e., has a $\mathbf{Z}$-base permuted by π. This can be made more precise by using the following ideas of Colliot-Thélène and Sansuc (1977).

Let $\mathscr{L}_\pi$ be the category of $\mathbf{Z}\pi$-modules which are finitely generated and free as abelian groups. Let $\mathscr{P}_\pi \subset \mathscr{L}_\pi$ be the subcategory of permutation modules and define $\mathscr{F}_\pi \subset \mathscr{L}_\pi$ to consist of those $F \in \mathscr{L}_\pi$ such that all $0 \to P \to E \to F \to 0$ with $P \in \mathscr{P}_\pi$ split. We say $F_1, F_2 \in \mathscr{F}_\pi$ are stably equivalent if $F_1 \oplus S_1 \approx F_2 \oplus S_2$ with $S_1, S_2 \in \mathscr{P}_\pi$. The stable equivalence classes form a monoid F_π under $\oplus$. If $M \in \mathscr{L}_\pi$, there is a resolution $0 \to M \to P \to F \to 0$ with $P \in \mathscr{P}_\pi$, $F \in \mathscr{F}_\pi$. We define $\rho(M) \in F_\pi$ to be the stable equivalence class of F. It can be shown that $\rho(M)$ is well defined.

Lemma 6.6. $\rho(M) = \rho(N)$ if and only if there are short exact sequences

$$0 \to M \to E \to P \to 0$$

$$0 \to N \to E \to Q \to 0$$

with $P, Q \in \mathscr{P}_\pi$.

For the proof, which is related to that of Schanuel's lemma, see Colliot-Thélène and Sansuc (1977).

Galois Theory

Corollary 6.7 $\rho(A*/k*) \in F_\pi$ is a birational invariant of K/k.

I shall denote this by $\rho(K)$. Note that $\rho(Q(k'/k, M)) = \rho(M)$ since we can choose $A = k'[M]$ here.

One can derive more easily manageable invariants from this as follows. Let R be a Dedekind ring and $\theta: \mathbf{Z}\pi \to R$. If $M \in \mathscr{L}_\pi$ let $(R \otimes_\pi M)_0 = (R \otimes_\pi M)/\text{torsion}$. If $M = \mathbf{Z}\pi/\pi'$ then $R \otimes_\pi M$ is a quotient of R and so is R or torsion. It follows that $(R \otimes_\pi P)_0$ is free for $P \in \mathscr{P}_\pi$, so we can define a homomorphism $c_\theta: F_\pi \to C(R)$ by sending F to the ideal class of the module $(R \otimes_\pi F)_0$ in the class group $C(R)$.

The example of $G = \mathbf{Z}/47\mathbf{Z}$ that I considered (Swan 1969) can be treated by the above method. One gets an ideal class in $C(\mathbf{Z}[\zeta_{47}])$ which is easily seen to be nontrivial.

The invariants c_θ do not suffice in the case $G = \mathbf{Z}/8\mathbf{Z}$ discussed in Section 3. In this case we can use the following argument of Lenstra (1974). If $Q(k'/k, M)$ is rational we have $0 \to M \to S_1 \to S_2 \to 0$ with $S_1, S_2 \in \mathscr{P}_\pi$. If it happens that $M^* = \text{Hom}_\mathbf{Z}(M, \mathbf{Z}) \in \mathscr{F}_\pi$, this sequence will split, so M is a direct summand of $S_1 \in \mathscr{P}_\pi$. This forces $H^1(\pi, M) = H^1(\pi, M^*) = 0$, which imposes nontrivial conditions on M.

7. STABLE EQUIVALENCE

Two field extensions F/k, F'/k are called stably birationally equivalent if we can adjoin indeterminates to make $F(x_1, \ldots, x_n) \approx F'(y_1, \ldots, y_m)$ over k. Zariski's cancellation problem asks whether this implies that $F' \approx F$ over k. Very little is known about this question.

The invariant ρ of Section 6 is a stable birational invariant since if we have a ring $A \subset K$ with the properties required in Section 6, we can use $A[x_1, \ldots, x_n] \subset K[x_1, \ldots, x_n]$. For fields of the form $Q(k'/k, M)$, it completely determines the stable isomorphism class. This is proved by Colliot-Thélène and Sansuc (1977) using ideas of Lenstra (1974).

Theorem 6.8 $Q(k'/k, M)$ and $Q(k'/k, N)$ are stably birationally equivalent if and only if $\rho(M) = \rho(N)$ in F_π.

Proof: By Lemma 6.6 it will suffice to show that $Q(k'/k, Y)$ and $Q(k'/k, X \times S)$ are birationally equivalent when $0 \to X \to Y \xrightarrow{j} S \to 0$ with S a permutation module. In fact, $Q(k'/k, X \times S) = Q(K/k, S)$ where $K = Q(k'/k, X)$ and we can apply the remark at the end of Section 5. Now in $Q(k'/k, Y) \supset K$ we have $1 \to K^* \to K^*Y \xrightarrow{\eta} S \to 0$ where $\eta(\alpha y) = j(y)$.

This sequence splits by Shapiro's lemma and Hilbert's Theorem 90, so $K*Y \approx K* \times S \supset X \times S$ and the required result follows easily.

8. LENSTRA'S THEOREM

In the case where G is an abelian group acting simply transitively on indeterminates $x_1, \ldots, x_n$, Lenstra (1974) has given a complete answer to the question of which $k_G = k(x_1, \ldots, x_n)^G$ are rational. Earlier partial results in this direction were found by Endo and Miyata (1973, 1973-1974, 1974) and Voskresenskii (1970a,b; 1971; 1973; 1974a,b). This happens if and only if the invariant ρ is trivial. (If char $k \mid |G|$, the definition of ρ must be modified.) In particular, if such fields are stably rational they are rational. Lenstra (1974) expresses his result in a very concrete form. The idea discussed at the end of Section 6 leads to the following necessary condition.

Lemma 6.9 (Lenstra 1974)   Let e be the exponent of G. Assume char $k \neq 2$ and let r be the largest power of 2 dividing e. If k_G is rational then $k(\zeta_r)/k$ is cyclic.

It is interesting to note that this condition plays a key role in the proof of Grunwald's theorem (Wang 1950) and appears again in Saltman's paper (1980) as a sufficient condition for the existence of a generic Galois extension.

The remaining obstructions to the rationality of k_G are ideal classes of the type discussed in Section 6. They are given explicitly by Lenstra (1974) and, in a somewhat different form, by Kervaire (1973-1974).

9. TORI

Voskresenskii (1970a,b; 1971, 1973; 1974a,b) pointed out that fields of the form $Q(k'/k, M)$ are precisely the function fields of algebraic tori. He used geometrically defined properties of tori to produce invariants for such fields. See also Endo and Miyata (1973, 1973-1974, 1974). This work was continued by Colliot-Thélène and Sansuc (1977). The discussion in Section 6 above follows closely the one in that paper. The main object of Colliot-Thélène and Sansuc (1977) is to study the relation of R-equivalence, the equivalence relation on the rational point $T(k)$ of T defined as follows: $a, b \in T(k)$ are directly R-equivalent if there is a morphism $\varphi : U \to T$ (over k) with U a rational variety such that $a, b \in \varphi(U(k))$. One of the main results of Colliot-Thélène and Sansuc (1977) is that R-equivalence on a torus is the same as rational equivalence and that the group $T(k)/R$ of equivalence classes is completely determined by the invariant $\rho(M)$.

10. STEENROD'S PROBLEM

My own interest in Noether's problem resulted from reading Masuda's paper (1968) at a time when I was looking at Steenrod's problem in topology. This problem runs as follows: We are given a finite group π, a finitely generated $\mathbf{Z}\pi$-module M, and an integer $n \geqslant 1$. Is there a finite complex X with π-action such that $H_0(X) = \mathbf{Z}$, $H_i(X) = 0$ for $i \neq 0, n$, and $H_n(X) = M$ as π-module? In my paper of 1969 I showed that the answer was negative for $\pi = \mathbf{Z}/23\mathbf{Z}$, $M = \mathbf{Z}/47\mathbf{Z}$ with π acting through $\pi \hookrightarrow F_{47}{}^*$ (Swan 1969). D. Kahn pointed out that it would suffice to know that a certain ideal of $\mathbf{Z}\pi$ was principal. Exactly the same question was raised by Masuda in connection with Noether's problem.

A number of interesting results on Steenrod's problem have recently been obtained by J. Arnold and G. Carlsson.

REFERENCES

Artin, E. (1944). *Galois Theory*, University of Notre Dame.
Chase, S., D. Harrison, and A. Rosenberg (1969). *Galois Theory and Cohomology of Commutative Rings*, Memoirs AMS 52.
Colliot-Thélène, J-L., and J-J. Sansuc (1977). La R-équivalence sur les tores, *Ann. Sci. E.N.S. 101*, 175-230.
De Meyer, F., and E. Ingraham (1971), *Separable Algebras over Commutative Rings*, Lect. Notes in Math., Springer-Verlag, Berlin.
Endo, S., and T. Miyata (1973). Invariants of finite abelian groups, *J. Math. Soc. Japan 25*, 7-26.
———— (1973-1974). Quasi-permutation modules over finite groups I, II. *J. Math. Soc. Japan 25*, 397-421 and *26*, 698-713.
———— (1974). On a classification of the function field of algebraic tori, *Nagoya Math. J. 56*, 85-104.
Fischer, E. (1915). Die Isomorphie der Invariantenkorper der endlichen Abelschen Gruppen linearer Transformationen. *Gott. Nachr.*, 77-80.
———— (1916). Zür Theorie der endlichen Abelschen Gruppen, *Math. Ann.* 77, 81-88.
Fried, E., and J. Kollar (1978). Automorphism groups of algebraic number fields. *Math. Z. 163*, 121-123.
Hilbert, D. (1892). Über die Irreduzibilitat ganzer rationaler Funktionen mit ganzzahligen Koeffizienten, *J. f. reine angew. Math. 110*, 104-129.

Kervaire, M. (1973-1974). *Fractions rationneles invariantes (d'après H. Lenstra)*, Sem. Bourbaki, exp. 445, Paris.

Kuyk, W. (1964). On a theorem of Emmy Noether, *Proc. Nederl. Akad. Wetensch. Ser. A,* 67, 32-39.

Lang, S. (1962). *Diophantine Geometry,* Interscience, New York.

Lenstra, H.W., Jr. (1974). Rational functions invariant under a finite abelian group, *Invent. Math. 215,* 299-325.

Masuda, M. (1968). Application of the theory of the group of classes of projective modules to the existence problem of independent parameters of invariants, *J. Math. Soc. Japan 20,* 223-232.

Saltman, D. (1980). Generic Galois extensions, *Proc. Nat. Acad. Sci. USA 77,* 1250-1251.

Šafarevič, I. (1954). Constructions of fields of algebraic numbers with given solvable Galois group, *Izv. Akad. Nauk SSSR Ser. Mat. 18,* 515-578.

Speiser, A. (1919). Zahltheoretische Satze aus der Gruppentheorie, *Math. Z. 5,* 1-6.

Swan, R.G. (1969). Invariant rational functions and a problem of Steenrod, *Invent. Math. 7,* 148-158.

Voskresenskii, V.E. (1970a). Birational properties of linear algebraic groups, *Izv. Akad. Nauk SSSR Ser. Mat. 34,* 3-19 [= *Math. USSR Izv. 4* (1970), 1-17].

———(1970b). On the question of the structure of the subfield of invariants of a cyclic group of automorphisms of the field $Q(x_1, \ldots, x_n)$, *Izv. Akad. Nauk SSSR Ser. Mat. 34,* 366-375 [= *Math. USSR Izv. 4* (1970), 371-380].

———(1971). Rationality of certain algebraic tori, *Izv. Akad. Nauk SSSR Ser. Mat. 35,* 1037-1046 [= *Math. USSR Izv. 5* (1971), 1049-1056].

———(1973). Fields of invariants of abelian groups, *Uspehi Mat. Nauk 28,* 77-102 [= *Russ. Mat. Surveys 28* (1973). 79-105].

———(1974a). Stable equivalence of algebraic tori, *Izv. Akad. Nauk SSSR Ser. Mat. 38,* 3-10 [= *Math. USSR Izv. 8* (1974). 1-7].

———(1974b). Some problems in the birational geometry of algebraic tori (in Russian), *Proc. Int. Congr. Math.,* Vancouver, 343-347.

van der Waerden, B. L. (1953). *Modern Algebra I,* F. Ungar, New York.

Wang, S. (1948). A counterexample to Grunwald's theorem, *Ann. Math. 49,* 1008-1009.

———(1950). On Grunwald's theorem, *Ann. Math. 51,* 471-484.

7

The Calculus of Variations

E. J. McShane

The calculus of variations began with problems of finding among all the (smooth) functions $x \mapsto y(x)$ that join two fixed points the one that gives the least value to some integral of the form

$$\int f(x, y, y') \, dx \qquad (7.1)$$

There was no unity in the methods of treatment of the separate problems. Lagrange observed that if a curve $C_0 : x \mapsto y(x)$ ($a \leq x \leq b$) minimizes the integral (7.1) in the class of curves with assigned endpoints, and a family of curves

$$C_\alpha : x \mapsto y(x, \alpha) \qquad a(\alpha) \leq x \leq b(\alpha), -\epsilon < \alpha < \epsilon$$

contains C_α for $\alpha = 0$ (we omit mention of banal continuity conditions), then if the C_α all satisfy the end-conditions, the function

$$J[C_\alpha] = \int_{a(\alpha)}^{b(\alpha)} f\left(x, y(x, \alpha), \frac{dy(x, \alpha)}{dx}\right) dx \qquad (7.2)$$

will have derivative 0 at $\alpha = 0$, since it has its least value there. A curve C_0 such that $dJ[C_\alpha]/d\alpha$ vanishes at 0 for all such families C_α is called *stationary*. The concept extends obviously to more complicated integrals, in which there may be more than one function y and more than one independent variable x, and derivatives of order higher than 1 may appear as arguments of f. From the property of being stationary Lagrange deduced differential equations, usually called the Euler-Lagrange equations, that y must satisfy. When these differential equations are satisfied, the function is stationary, not necessarily minimizing. But for applications to mechanics the property of being stationary is outstandingly important. If a mechanical system moves in accordance with Newton's laws, its state at each time t is specified by a vector-valued function $t \mapsto y(t)$ that is a stationary function for a certain integral. This allows a most useful freedom of choice of coordinate system, and became and continues to be indispensable in the study of mechanics.

Emmy Noether's contribution to the calculus of variations consists of two theorems about solutions of the Euler-Lagrange equations, for both simple- and multiple-integral problems. In brief summary, she showed that if the integral is invariant under a group of mappings of functions into functions, the stationary functions satisfy corresponding restrictive conditions, which in the case of simple-integral problems take the form of first integrals of the Euler-Lagrange equations. Such first integrals include the conservation laws of physics, and are therefore of special interest to physicists. Perhaps Emmy Noether's attention was drawn to such conservation laws by her physicist brother. It did not occur to me to ask her about this when I used to meet her, in the early 1930s; in fact, if it had occurred to me, I probably would have lacked the presumption of asking such a question of a mathematician so much older and more distinguished than I, even though that mathematician was the friendly and approachable Emmy Noether.

The case of simple integrands involving only the first derivative of the n-vector-valued function $y = (y^{(1)}, \ldots, y^{(n)})$ is much the simplest to discuss, and is still general enough to include all the conservation laws of physics; in fact, I know of no book on the calculus of variations that indicates that "Noether's theorem" goes beyond this. So we suppose that $(x,y,r) \mapsto f(x,y,r)$ is defined for all (x,y) in an open set G in $(n+1)$-space and all n-tuples r. Then to each curve

$$C: x \mapsto y(x) \qquad a \leq x \leq b$$

with all its points $(x, y(x))$ in G there corresponds a number

$$J[C] = \int_a^b f(x, y(x), y'(x))\, dx \tag{7.3}$$

If a curve C has a representation

$$t \mapsto (X(t), Y^1(t), \ldots, Y^n(t)) \qquad c \leq t \leq e$$

in which dX/dt is positive, it also has a representation

$$x \mapsto y(x) \qquad X(c) \leq x \leq X(e) \tag{7.4}$$

The Calculus of Variations

and $J[C]$ can be computed by (7.3). But it is convenient to be able to compute $J[C]$ from the given representation. So for each z in G and each $(n+1)$-tuple $p = (p^0, \ldots, p^n)$ with $p^0 > 0$ we define

$$F(z,p) = f\left(z^0, z^1, \ldots, z^n, \frac{p^1}{p^0}, \ldots, \frac{p^n}{p^0}\right) p^0 \tag{7.5}$$

If we write $(z^0(t), \ldots, z^n(t))$ for $(X(t), Y^1(t), \ldots, Y^n(t))$, by change of variable,

$$J[C] = \int_c^e F\left(z(t), \frac{dz(t)}{dt}\right) dt \tag{7.6}$$

Suppose now that C is a curve in G with nonparametric representation (7.4); we write a, b for $X(c), X(e)$. For each α in an interval $(-\epsilon, \epsilon)$ let C_α be a curve with parametric representation

$$t \mapsto z(t, \alpha) \qquad c \leq t \leq e$$

where all z^i are continuously differentiable and C_0 is C. If we write

$$\eta^i(t) = \left.\frac{\partial z(t, \alpha)}{\partial \alpha}\right|_{\alpha=0} \tag{7.7}$$

by integration by parts we obtain

$$\frac{d}{d\alpha} J[C_\alpha]\Big|_{\alpha=0} = \int_c^e \sum_{j=0}^n \left[F_{z^j}(z(t,0), z'(t,0))\right.$$

$$\left. - \frac{d}{dt} F_{p^j}(z(t,0), z'(t,0))\right] \eta^j(t)\, dt$$

$$+ \sum_{j=0}^n \left[F_{p^j}(z(e,0), z'(e,0)) \eta^j(e)\right.$$

$$\left. - F_{p^j}(z(c,0), z'(c,0)) \eta^j(0)\right] \tag{7.8}$$

In particular, if we choose any number j in the set $\{0, \ldots, n\}$ and any smooth function η vanishing at c and e, then if C_0 is stationary, equation (7.8) reduces to

$$0 = \int_c^e \left[F_{z^j}(z(t,0), z'(t,0)) - \frac{d}{dt} F_{p^j}(z(t,0), z'(t,0))\right] \eta(t)\, dt$$

and since this holds for all such η the expression in brackets must be identically 0. If we choose the particular representation $x \mapsto y(x)$ for C_0, as we may, this yields by (7.5) the $n+1$ equations

$$f_x(x,y(x),y'(x)) - \frac{d}{dx}\left[f(x,y(x),y'(x)) - \sum_{i=1}^{n} y^{i'}(x) f_{r^i}(x,y(x),y'(x))\right] = 0 \qquad (7.9)$$

$$f_{y^i}(x,y(x),y'(x)) - \frac{d}{dx} f_{r^i}(x,y(x),y'(x)) = 0$$

These are the Euler-Lagrange equations.

Suppose next that for each α in $(-\epsilon,\epsilon)$ the transformations

$$z^j \mapsto \Phi^j(z,r,\alpha) \qquad (7.10)$$

are defined and continuously differentiable for z in G and r in n-space, and

$$\Phi^j(z,r,0) = z^j$$

If C_0 is any curve in G with nonparametric representation $x \mapsto y(x)$, $a \leq x \leq b$, for each α in $(-\epsilon,\epsilon)$ the functions

$$t \mapsto \Phi(t,y(t),y'(t),\alpha) \qquad a \leq t \leq b$$

define parametrically a curve C_α; and if the interval $(-\epsilon,\epsilon)$ is small enough, as we shall suppose, C_α is in G and can be nonparametrically represented. So $J[C_\alpha]$ is defined. If it has the same value for all α in $(-\epsilon,\epsilon)$, the integral J along C_α is said to be invariant under the set of transformations (7.10).

It is now easy to prove a special case of Emmy Noether's first theorem:

If C_0 is a stationary curve for the integral (7.3) and the integral along each subarc of C_0 is stationary under transformation (7.10), then for $a \leq x \leq b$,

$$\sum_{i=1}^{n} f_{r^i}(x,y(x),y'(x))\, \Phi_\alpha^i(x,y(x),y'(x),0)$$

$$+ \left[f(x,y(x),y'(x)) - \sum_{i=1}^{n} y^{i'}(x) f_{r^i}(x,y(x),y'(x))\right] \Phi_\alpha^0(x,y(x),y'(x),0)$$

$$= \text{constant} \qquad (7.11)$$

The Calculus of Variations

With C_α defined as above, we obtain from (7.7)

$$\eta^j(t) = \Phi^j_\alpha(t, y(t), y'(t, 0)) \qquad j = 0, 1, \ldots, n \tag{7.12}$$

For each subinterval $[c, e]$ of $[a, b]$ we write (7.8) for the arc $x \mapsto y(x)$, $c \leqslant x \leqslant e$. The left member is 0 because of the invariance of the integral, and by (7.9) the integrand of the integral in the right member is 0 because C_0 is a stationary curve. So (7.8) reduces to

$$0 = \int_c^e \frac{d}{dt} \sum_{j=0}^n \left[F_{p^j}(z(e,0), z'(e,0))\, \eta^j(t) \right] dt \tag{7.13}$$

Since this holds for all (c, e) contained in (a, b), the integrand is 0. The representation of C_0 is nonparametric, with $z^0(t) = t$, so by (7.5) and (7.12),

$$\frac{d}{dt}\left\{ \left[f - \sum_{i=1}^n y^{i'}(x) f_{r^i} \right] \Phi^0_\alpha + \sum_{i=1}^n f_{r^i} \Phi^i_\alpha \right\} = 0 \tag{7.14}$$

the functions being evaluated at $(x, y(x), y'(x))$. This implies (7.11).

The generalization of this theorem to integrands dependent on higher derivatives of the y^i is not essentially different, though tedious. If f depends on $x, y(x)$, and the derivatives of order up to mth, we can define a function F of $m + 1$ ordered $(n + 1)$-tuples, of which (7.5) is a special case, that allows change of representation of the curves C_α. Integration by parts that gets rid of the derivative of the η^i in the integrand leaves a function of the endpoints that generalizes the end-terms in (7.8) and remains constant along C_0 if the integral over each subarc is invariant under (7.10), or under the more general transformations whose right members are functions $\Phi^j(z, r_{[1]}, \ldots, r_{[m]}, \alpha)$ of m n-tuples $r_{[h]}$.

For the generalization to multiple integrals, in which $x = (x^1, \ldots, x^m)$ and the integral is over a set Δ in m-space, it is enough to consider sets Δ that are parallelepipeds (cartesian products of m intervals in one-space). Each integration by parts is with respect to one variable; the integral is written as an iterated integral, and the "integrated part" appears as the difference of the integrals of an easily computed function over two opposite faces $\{x^{(\alpha)} = c^{(\alpha)}\}$ and $\{x^{(j)} = e^{(j)}\}$ of Δ. This difference can be written as in (7.12) as the integral over Δ of the partial derivative of the expression with respect to $x^{(\alpha)}$, and the whole integrated part [generalizing the last sum in (7.8)] is the integral over Δ of the divergence of a computable but complicated expression. Conclusion (7.11) of the special case takes the form that this divergence is identically 0, as in (7.14).

The generalization to transformations (7.10) dependent on several parameters α is trivial. But there is a more general type of transformation. The transformations form a *continuous group* if they are a group whose members correspond to a finite set of arbitrary functions and finitely many derivatives of those functions. For these Emmy Noether proved her second theorem:

If the integral along every arc C_0 is invariant under the transformations of a group whose members are determined by ρ arbitrary functions and their derivatives of order up to σ, there exist ρ identical relationships between the Lagrangians [the left members of (7.9)] and their derivatives of order $\leq \sigma$. The converse is also true.

The proof is closely related to that of the first theorem, but we shall not attempt to sketch it.

Emmy Noether's theorems in the calculus of variations serve to unify knowledge about conservation laws; they have applications in classical mechanics and in relativity theory; and they form a linkage between the calculus of variations and Lie group theory. Yet they are ignored in many books on the calculus of variations and touched lightly in the others. They have fallen victim to a change in fashions. They constitute a major contribution to a highly formalistic aspect of the calculus of variations that received much attention in the nineteenth century, but was already becoming less interesting to analysts when her theorems appeared, in Ref. 13.* Problems involving multiple integrals and higher derivatives are largely ignored today, so it is quite natural that of her theorems, only the simple special case that we have presented still survives in the literature.

---

*Numbered references are to Noether's papers listed at the end of this book.

8

Commutative Ring Theory

Robert Gilmer

Commutative ring theory was the primary focus of Emmy Noether's work during the approximate period 1920-1926. During this time, she published five papers [19, 24, 25, 30, 31]* and three abstracts [26, 27, 28] in what was called, at the time, "the general theory of ideals." Also, two students—Grete Hermann and Heinrich Grell—completed dissertations in commutative ring theory under Noether's direction during this period, and a third student, Rudolf Hölzer, died before his work in the area was completed (Dick 1970, p. 42). The year 1920 also marks a turning point in Emmy Noether's career. It was then that she began to be recognized as a mathematician in her own right, as opposed to an assistant of Hilbert and Klein. By 1923, she was recognized as the leader of the Göttingen School of Algebra, a position she maintained until her departure from Göttingen in 1933 (Kimberling 1972, pp. 145, 146). Her monumental paper *Idealtheorie in Ringbereiche*, published in 1921, was responsible for much of the recognition she received; its elegant, axiomatic approach was novel in the area at the time, but more than anything else, she gained the attention and respect of the mathematical community because of the results she obtained. For Noether did more than generalize results known for polynomial rings to rings satisfying the ascending chain condition for ideals; in addition to that, she proved significant new results that had not been proved for polynomial rings. Her results

---

*Numbered references are to Noether's papers listed at the end of this book.

tangibly indicated the power of the axiomatic approach and the wealth of good mathematics it was to yield.

In discussing Noether's work in commutative ring theory, we concentrate on Ref. 19, which we refer to briefly as *Idealtheorie,* and on the 1927 paper *Abstrakter Aufbau der Idealtheorie in algebraischen Zahl- und Funktionenkörpern* [31], which we abbreviate in the text as *Aufbau.* Mathematicians outside Germany came to know of the results of these two papers primarily through Volume II of van der Waerden's *Moderne Algebra* (1931). [For a timely account of the historical impact of the work, see Dieudonné (1970, pp. 136-7).] In fact, Chapter 12 of *Moderne Algebra* presents a refined version of the main results of *Idealtheorie,* while Chapter 14 is a fairly faithful account of *Aufbau.* Outside *Idealtheorie* and *Aufbau,* Noether's main contribution to the area is probably the form of the Normalization Lemma proved in Ref. 30. As is well known, algebraic geometry and algebraic number theory are the two main roots from which modern commutative algebra has sprung. This being so, it seems highly appropriate that the historical basis for *Idealtheorie* was the work of Hilbert, Lasker, and Macaulay on ideals of polynomial rings, their corresponding algebraic varieties, and their decomposition into primary components, whereas Richard Dedekind's work on the ideal theory of rings of algebraic integers of a finite algebraic number field provides the primary motivation for *Aufbau.*

Before discussing *Idealtheorie* and *Aufbau* separately in some detail, we make a few general remarks that apply to each of them. To some degree, the aspects we mention may be a reflection of a time when demands on journals' space were not as great, and when printing costs were not so significant a factor in determining how a published paper was presented. The pace of the two papers is leisurely, but not dawdling; they contain approximately 25 references, and yet each is essentially self-contained. The proofs are detailed enough for a reasonably informed reader to follow without much head scratching. As was common for papers published at the time, footnotes abound—there are 51 in *Idealtheorie* and 34 in *Aufbau.* References and examples are frequently consigned to the footnotes. Also, the genesis of results, references to similar methods or considerations, and limitations on the theory are mentioned in the footnotes. Much more than the text proper, the footnotes of *Idealtheorie* and *Aufbau* indicate certain personal characteristics of Emmy Noether—a generosity in giving credit to others, a corresponding modesty about the magnitude of her own contributions, and a profound knowledge of, and admiration for, the work of Richard Dedekind. In short, these papers are a pleasure to read, and they seem to provide more insight and motivation than the typical paper published today.

IDEALTHEORIE IN RINGBEREICHEN

Hilbert (1890) proved that each ideal of the polynomial ring in n variables over the complex field is finitely generated, and his complex Nullstellensatz is contained in an 1893 paper (Hilbert 1893). [Steinitz's paper (1910) represents the birth of abstract field theory.] E. Lasker (1905) proved that each ideal of $C[X_1, \ldots, X_n]$ or of $Z[X_1, \ldots, X_n]$ is a finite intersection of primary ideals, and F. S. Macaulay (1913) gave an algorithmic process for determining a primary representation of an ideal A of such a polynomial ring, given a finite set of generators of A. As stated in the introduction, these four papers (Hilbert 1890, 1893; Lasker 1905; Macaulay 1913) provided the main motivation for *Idealtheorie*. Of course, a major step that Noether took was to move from the concrete—polynomial rings or Dedekind's "Ordnungen" of algebraic numbers (Dirichlet 1894, § 170)—to an abstract algebraic structure (a commutative ring in this case). Ring theory itself was in an early stage of development at the time; the term "Ring" or "Zahlring" seems to have been first used by Hilbert (1894-1895), while Kronecker (1882, § 5) had previously used the term "Integritätsbereich" (literally, "domain of integrity"). A. Fraenkel (1914) first defined the concept of an abstract ring, but his definition is not the one in common usage today; the current definition of a (commutative) ring seems to have first appeared in a paper of M. Sono (1917). At any rate, the subject was in such an early state of development that Noether gave a proof in footnote 32 of *Idealtheorie* that a commutative ring with identity has a unique identity element.

The most important concept in *Idealtheorie* is, without doubt, that of the ascending chain condition (a.c.c.) for ideals. Noether proves the now-standard result that each ideal of the commutative ring R is finitely generated if and only if the a.c.c. for ideals is satisfied in R. [The term "noetherian ring" to describe a ring in which each ideal is finitely generated apparently was originated by C. Chevalley (1943).] As precedents for considering the chain condition, Noether cites Dedekind (Dirichlet 1894, VIII, p. 253) and Lasker (1905, p. 56); one of Noether's main contributions, then, was in isolating the condition, recognizing its utility, and defining it in an abstract setting. On the other hand, two other familiar equivalent forms of the chain condition—the maximal condition and the principle of divisor induction—are missing from *Idealtheorie*, and this causes some proofs of standard results in the paper to be longer than expected. A notable case in point is the proof that each ideal of a noetherian ring is a finite intersection of irreducible ideals, which covers more than a page in *Idealtheorie*. Noether gives an alternate proof in *Aufbau* in twelve lines, using in essence the principle of divisor induction.

Noether considers four different types of decomposition in Ref. 19. Of these four, primary decomposition is the second, as well as the type with the best endurance record. But in Ref. 19, Noether shows no preference for primary ideals over irreducible ideals, and without question, her first decomposition theorem is the result she uses with greatest effect. To describe this result, as well as the third and fourth decomposition theorems, we review some terminology from Ref. 19. The basic object of study in the paper is a commutative noetherian ring R, which may not have an identity element. (Noether remarks that the chain condition hypothesis on R is not needed in some of the results.) If $\{B_i\}_{i=1}^{k}$ is a finite family of ideals of R, then the intersection representation $\cap_{i=1}^{k} B_i$ is said to be a *shortest representation* if no B_i contains the intersection of the other ideals B_j, and the representation is said to be *reduced* if it is a shortest representation and no B_i can be replaced by a strictly larger ideal C_i to obtain the same intersection. If A and B are ideals of R, then A *is prime to* B (A is *relativprim zu* B in Ref. 19) if $B : A = \{r \in R : Ar \subseteq B\} = B$; these ideals are *mutually prime* if A is prime to B and B is prime to A, and an ideal C of R is *prime-irreducible* if C cannot be expressed in the form $A \cap B$, where A and B properly contain C and are mutually prime. Also, A and B are *comaximal* ("*teilerfremd*") if $A + B = R$, and an ideal C is *comaximal-irreducible* if C is not expressible as $A \cap B$, where A and B are comaximal ideals properly containing C. In §§ 2,3 of Ref. 19, Noether shows that each ideal of R admits a shortest representation as an intersection of irreducible ideals, that such a representation is reduced, and that the number of components appearing in such a representation is invariant; we refer to the result that each ideal A of R admits a reduced representation $A = \cap_{i=1}^{k} B_i$, with each B_i irreducible, as the first decomposition theorem. In §§ 4,5, Noether uses the first decomposition theorem to show that an irreducible ideal of R is primary; she then proves the invariance of the set of belonging prime ideals in either a shortest representation or a reduced representation of an ideal by greatest primary components. A proof of uniqueness of the isolated components in such representations is delayed until after the third decomposition theorem is proved in §6. Uniqueness of the set of belonging primes and isolated components in a shortest representation by greatest primary components are examples of significant results Noether obtained which were not included in earlier work of Lasker and Macaulay in the concrete setting of polynomial rings. The third decomposition theorem states that each ideal of R is uniquely expressible in the form $\cap_{i=1}^{k} A_i$, where the ideals A_i are prime-irreducible and A_i and A_j are mutually prime for $i \neq j$. While this theorem has seen little subsequent use, one of the results Noether establishes for its proof is familiar: the ideal B of

R is prime to the ideal A if and only if no belonging prime of B is contained in a belonging prime of A. The fourth decomposition theorem is proved for a noetherian ring with identity; its statement is the same as the third decomposition theorem, except that "prime-irreducible" and "mutually prime" are replaced by "comaximal-irreducible" and "comaximal." The fourth decomposition theorem is recorded by van der Waerden (1931, §85). It is equivalent to the statement that a nonzero noetherian ring with identity is uniquely expressible as a finite direct sum of nonzero indecomposable rings, and this equivalent version is the form in which the theorem is best known today.

In §9 of Ref. 19, Emmy Noether defines a (left) module A over a ring T that need not be commutative. She remarks that each submodule of A is finitely generated if and only if the a.c.c. for submodules is satisfied in A, and she proves that a finitely generated unitary module over a commutative noetherian ring with identity is a noetherian module. She states that the first decomposition theorem generalizes to modules, since its proof involves only the concepts of inclusion and intersection of ideals, and not the notion of an ideal product. On the other hand, products of ideals are used in the proofs of the other three decomposition theorems, and hence these theorems do not generalize to modules without further ado.

We close our discussion of *Idealtheorie* with some remarks about the proofs of the uniqueness of the belonging prime ideals and isolated primary components in a representation by greatest primary components, as these proofs appear in Ref. 19 and in van der Waerden's book (1931). In the latter, these uniqueness theorems are proved using the concept of the ideal quotient $A : B$, which W. Krull introduced (1922, p. 83). On the other hand, ideal quotients do not appear in *Idealtheorie* (in spite of the way we have defined the notion "A is prime to B"), and Noether's proofs of the uniqueness theorems in Ref. 19 are based on results related to her proof of the first decomposition theorem. The explanation for this is quite simple. Krull published a paper (1923) entitled *Ein neuer Beweis für die Hauptsätze der allgemeinen Idealtheorie*; the "main theorems of general ideal theory" referred to in the title are the uniqueness theorems, and the "new proofs" are proofs using ideal quotients—essentially the proofs that appear in van der Waerden (1931). In this connection, we remark that Krull basically refers to Ref. 19 for the proofs that (1) each ideal of a noetherian ring R is a finite intersection of irreducible ideals, and (2) that an irreducible ideal of R is primary; in other words, he proves uniqueness of the belonging primes, etc., on the basis of the existence of a representation as a finite intersection of primary ideals. The elementary proof of (1), using the principle of divisor induction, and the short, elegant proof of (2) that appear in van der Waerden (1931) stem from *Aufbau*.

ABSTRAKTER AUFBAU DER IDEALTHEORIE IN ALGEBRAISCHEN ZAHL- UND FUNKTIONENKÖRPERN

In the opening sentence of *Aufbau*, Emmy Noether indicates that the purpose of the paper is to provide an abstract chacterization of those rings whose ideal theory agrees with that of the ring of all algebraic integers of a finite algebraic number field—that is, whose ideals are uniquely expressible as finite products of prime ideals [Dirichlet (Dedekind's Suppl. XI)]. In this aim she largely succeeds, although Noether did not, in fact, establish a well-known characterization that has sometimes been attributed to her (Schmidt 1936, p. 443; Cohen 1950, p. 31). (We defer elaboration on this statement until later in this section.) To be more specific, Noether considers in the first eight sections of *Aufbau* a commutative ring R satisfying various combinations of the following five axioms [Noetherschen fünf Axiome (Kubo 1940)]:

I. R is noetherian.
II. R/A is artinian for each nonzero ideal A of R.
III. R has an identity element.
IV. R is an integral domain.
V. R is integrally closed in its quotient field.

In Satz VI of Ref. 31, she proves that if Axioms I-V are satisfied in R, then each nonzero proper ideal of R is uniquely expressible as a finite product of proper prime ideals of R, and nonzero proper primes of R are maximal. The form of the converse she proves in §9 of Ref. 31 is complicated to state because she does not hypothesize the existence of an identity element, but in the case of rings with identity, the result proved is the following. Assume that S is a commutative ring with identity such that each nonzero proper ideal of S is uniquely expressible as a finite product of proper prime ideals of S, that nonzero proper prime ideals of S are maximal, and that powers of each nonzero proper ideal of S properly descend. Then Axioms I-V are satisfied in S.

The concept of integral dependence is to *Aufbau* what the a.c.c. is to *Idealtheorie*—a notion occurring in Dedekind's work in algebraic number theory which Noether defined in an abstract setting and which has evolved into a concept of basic importance in commutative algebra. Noether's definition is the following. Assume that T is an integral domain with identity and R is a subring of T containing the identity element. An element t of T is *integral* over R if the ascending sequence $\{(t^0, t^1, \ldots, t^{i-1})\}_{i=1}^{\infty}$ of R-submodules of T becomes stationary; equivalently, t is integral over R if t satisfies a monic polynomial over R. In §1 of Ref. 31, Noether establishes what have come to be regarded as some basic properties of integral dependence—for example, integral dependence is transitive, and the elements of T integral over R form a subring of T. This section closes with two curious results labeled as Dedekind Folgerung I and Dedekind Folgerung II. In particular, II states that if D is a noetherian integrally closed domain with quotient field K and if $t \in K$ is not integral over D, then t admits a representation in the form m/n, where m^2/n is not integral over D; this result is subsequently used to prove that if P is a prime ideal of R, a ring satisfying Axioms I-V, then there exists no ideal of R properly between P and P^2.

An outline of Noether's proof that Axioms I-V imply unique factorization into prime ideals is fairly easy to sketch; results of *Idealtheorie* are used strongly in the proof. Thus, if R satisfies I-V, then each nonzero proper ideal of R admits a representation $\cap_{i=1}^{k} Q_i$ by greatest primary components. An elementary proof shows that Axiom II implies that nonzero proper prime ideals of R are maximal, so that the ideals Q_i are pairwise comaximal, and hence $\cap_{i=1}^{k} Q_i = Q_1 Q_2 \cdots Q_k$. To complete the proof of existence of a representation into a product of prime ideals, it suffices to show that primary ideals of R are prime powers; this is where Axiom V is used. As stated above, Noether shows that for P maximal in R, there exists no ideal of R properly between P and P^2. A known result of Sono (1913) then implies that for each positive integer k, the only ideals of R between P and P^k are $P, P^2, \ldots, P^k$, and this completes the proof of existence of factorization into prime ideals. Incidentally, consistent with Noether's practice of keeping matters as self-contained as possible, she provides an independent proof of Sono's result. As for uniqueness, since decomposition into greatest primary components and factorization into powers of distinct maximal ideals coincide in the presence of Axioms I-V, then uniqueness of the isolated primary components implies that to establish uniqueness of prime factorization, it is necessary to show only that $P^k \neq P^{k+1}$ for P a nonzero proper prime of R; Noether proves this in the expected way, using Cramer's rule and the fact that P^k is finitely generated.

To consider the proof in Ref. 31 of the converse, take a ring S as described in the introductory paragraph of this section. Thus, Axiom III is satisfied in S by hypothesis (as stated above, Noether did not restrict to rings with identity in her proof of the converse). Axioms I, II, and IV are derived easily —I and II follow from the fact that a nonzero ideal $A = M_1^{e_1} \cdots M_k^{e_k}$ of S has only finitely many divisors (namely, those ideals of the form $M_1^{f_1} \cdots M_k^{f_k}$, with $0 \leqslant f_i \leqslant e_i$ for each i), and IV follows from prime factorization and the fact that $A^k \neq A^{k+1}$ for a nonzero proper ideal A of S. Thus, it remains to prove that S is integrally closed. In the introduction of *Aufbau*, Noether indicates that Dedekind was aware that integral closure follows from representation through prime powers; for this statement she refers to a remark in §172, page 522, of the third edition of Dirichlet's book (1894). We do not have access to the third edition, and have found no such remark in the fourth edition, but it is certainly the case that several of the concepts and results in Noether's proof that S is integrally closed have analogues in Supplement XI of the fourth edition. Rather than sketch the proof per se, we mention some of its component results and ideas. The first step of the proof shows that S/A is a principal ideal ring for each nonzero ideal A of S; the notions of multiplication ideal and invertible ideal are implicit in the proof; fractional ideals of S are defined, and it is shown that the set of nonzero fractional ideals of S forms a group under multiplication; and finally, the approach of considering P/P^2 as a vector space over S/P in order to determine the structure of the set of ideals between P and P^2 is used in the proof. Each of the items mentioned in the preceding sentence is instantly recognized by persons working in commutative algebra today as standard and significant—a tribute to Emmy Noether's lasting influence in the area.

In §3 of *Aufbau* Noether proves that if D is a domain satisfying I-V, then these axioms are also satisfied in the integral closure of D in a finite separable extension of the quotient field of D. Since each principal ideal domain satisfies I-V, Dedekind's result on unique factorization of ideals of finite algebraic number fields then follows from results of Ref. 31, but more is obtained as well. In particular, if K is an algebraic function field of one variable over a finite field or a field of characteristic 0, then each integrally closed subring of K containing the prime subfield admits unique factorization into prime ideals. As an additional application of the results, Noether proves (in current terminology) that the Kronecker function ring $D[Y]^v$ of $D[Y]$, where $D = k[X_1, \ldots, X_n]$, is a principal ideal domain, and hence satisfies I-V.

We return to the statement made in the introduction concerning a characterization of rings with unique factorization of ideals attributed to Noether. I. S. Cohen (1950) used the term *Dedekind domain* to define an integral domain with identity in which each nonzero proper ideal is a finite product of proper prime ideals; a result of Matusita (1944) shows that in a Dedekind domain factorization into proper prime ideals is unique. A well-known theorem states that an integral domain D with identity is a Dedekind domain if and only if Axioms I, V, and the following Axiom II' are satisfied in D.

(II)' Each nonzero proper prime ideal of D is maximal.

Using the fact that a commutative ring R with identity is artinian if and only if R is noetherian and proper primes of R are maximal (Akizuki 1935; Cohen 1950), it follows that I and II are equivalent to I and II$'$, and I, II$'$, and V have sometimes been referred to as "Noether's three axioms" (Gilbert and Butts 1968). While Axioms I, II, and V characterize Dedekind domains among integral domains with identity, this characterization—attributed to Noether by Schmidt (1936)—is, in fact, not contained in Ref. 31. This would remain true even if uniqueness of prime ideal factorization were taken as part of the definition of a Dedekind domain. While Noether proved that I, II, and V imply unique factorization into products of maximal ideals, in the reverse direction, she assumes that nonzero proper primes are maximal in proving that unique factorization implies I, II, and V. Kubo (1940) proved that if each nonzero proper ideal of the integral domain D with identity is uniquely expressible as a finite product of proper prime ideals, then conditions I, II, and V are satisfied in D.

We conclude this section with some general remarks about *Aufbau*. Certain contrasts with *Idealtheorie* are apparent. Noether writes with more confidence in *Aufbau*; one gathers that she feels less need to explain or justify her methods or results—the axiomatic approach has gained some acceptance, and certain notions about commutative rings have become old hat and need little elaboration. On the other hand, her admiration of Dedekind's work and deference to him come through more strongly in *Aufbau*. There are at least six references in the paper to Dirichlet (1894) or to the third edition of that book. One of these references occurs in footnote 10, where Noether states that Dedekind already knew for number fields that a finitely generated module over a noetherian ring is a noetherian module (Dirichlet 1894), and that this proof of Dedekind represents the first use of chain conditions in the literature. It's as if Noether is giving Dedekind his due for the acclaim she has received for her use of the ascending chain condition—the repetition of an act she has already performed in *Idealtheorie*. Auguste Dick has written that Noether urged her students to become familiar with all editions of Dirichlet (1894), and that one of Noether's favorite expressions was "Alles steht schon bei Dedekind" (Dick 1970, p. 25). Indeed, results of *Aufbau* indicate that Dedekind's work was a veritable treasure field for a person with Emmy Noether's powers of discernment.

EMMY NOETHER'S INFLUENCE IN COMMUTATIVE RING THEORY

Perhaps more than any other person, Emmy Noether is identified with the axiomatic approach in mathematics, but this is particularly true in algebra. The power of this method requires no attestation. Concerning Noether's influence on modern algebra in general, we refer to an article of Garrett Birkhoff (1976); see also Kaplansky (1973).

One way to measure Emmy Noether's influence in commutative ring theory is in terms of topics persons in the area have worked on since *Idealtheorie* appeared. For almost twenty years, until almost 1940, topics in or closely related to *Idealtheorie* dominated research in the area; these included various chain conditions, studies of different types of primary ideals and characterizations of rings that admit primary decomposition, development of the theory of noetherian rings, investigations of rings (possibly with zero divisors) in which each ideal of a certain class is representable as a product of prime ideals, and work in determining the structure of multiplication rings (rings in which each ideal is a multiplication ideal). Some workers stuck with integral domains with identity or with a more concrete setting, others worked in rings with identity, and a few retained the generality of an arbitrary commutative ring; almost without fail, the noetherian case was a model or an interesting test of the work in question. Some work directly related to Refs. 19 and 31, such as the papers of Matusita (1944), and Cohen (1950) already cited, continued beyond 1940, and noetherian ring theory remains an active topic today. There were, of course, other developments in commutative ring theory during the period 1920-1940 that had little connection with *Idealtheorie* or *Aufbau*; some examples that come to mind are the work of Grell on regular quotient rings, the work of Arnold, van der Waerden, Prüfer, Krull, and Lorenzen on systems of ideals, the work of Krull and others on general valuations, and Krull's work on power series rings, dimensions theory of local rings, and his *Beiträge* series. However, in this list of names, at least three persons—Grell, van der Waerden, and Krull—studied with Emmy Noether. This brings us to a second aspect of her influence.

Göttingen was a center for algebra from approximately 1920 until the time of Emmy Noether's departure in 1933. She was very much involved in the work of the mathematics institute, its seminars, and other activities. Algebraists from all over the world came to Göttingen to work with her, and by all accounts she gave of her time and energy freely to students and visitors. Also, she participated actively in the mathematical community outside Göttingen, attending and speaking at meetings of the Deutschen Mathematiker-Vereinigung, visiting other universities, etc. The point is that Emmy Noether's sphere of influence was enlarged by the large number of contacts she made. Among persons who came to Göttingen to work with Noether, we select three for additional comment. Listed in order of the time when they were in Göttingen, these persons are Wolfgang Krull, B. L. van der Waerden, and Shinziro Mori.

Wolfgang Krull was the leading developer, contributor, and innovator in commutative ring theory during the period 1920-1940. He wrote more than 50 strong research papers in the area during this time. A number of Krull's contributions outside topics considered in Refs. 19 and 31 have already been mentioned; one that has not been mentioned is his book (entitled *Idealtheorie*) published in the Ergebnisse series (1935). This monograph is a summary report of work in the general theory of ideals up to 1934; it remains an authoritative reference today on such work. Although Hermann Weyl (1935, p. 210), lists Krull as one of Noether's "pupils proper," Krull in fact wrote his thesis, entitled *Über Begleitmatrizen und Elementarteiltheorie,* under the direction of A. Loewy at the University of Freiburg (Breslau) in 1921. Krull spent the 1921-1922 academic year at Göttingen, and returned to the faculty of the University of Freiburg in the fall of 1922. Krull published his first complete paper in 1922, and it was not until his eighth paper (1925) that he returned to the topic of his thesis. Thus, it is clear that Emmy Noether had a strong and enduring influence on Krull's career.

As Weyl has written, van der Waerden came to Göttingen from Holland in 1924-1925 "as a more or less finished mathematician and with ideas of his own; but he learned from Emmy Noether the apparatus of notions and the kind of thinking that permitted him to formulate his ideas and solve his problems" (Weyl 1935, p. 210; see also Kimberling 1972, p. 145). Indeed, while van der Waerden did write a few papers in ideal theory, his main work was more in an algebro-geometric vein. But as all the world knows, Noether's main influence through van der Waerden is in the two volumes of *Moderne Algebra* (1931), whereby the mathematical community at large became students of E. Artin and Noether. The impact of this book and its reflection of Noether's methods and work have already been referred to; without doubt, *Moderne Algebra* (now in its seventh edition under the title *Algebra*) is the instrument through which Emmy Noether's influence has been greatest.

Shinziro Mori received his degree at Kyoto University in 1919; he was a student of M. Sono. Mori accepted a position as professor at Hiroshima Higher Normal College after his graduation, and he retained a profound interest in ideal theory during this time. In 1928 he traveled to Göttingen and pursued his study under Emmy Noether's direction for three years. In 1931 he accepted a position as professor at Hiroshima University, a post he retained until his retirement in 1956. Mori published more than 30 papers in the general theory of ideals during the ten-year period 1931-1940; most of these papers appear in *Journal of Science of Hiroshima University*. Noether's influence on Mori's work is clear; the topics he worked on—chain conditions, primary decomposition, factorization into prime ideals, and multiplication rings—were all suggested by Ref. 19 or Ref. 31. Moreover, Mori consistently chose as his basic object of study an arbitrary commutative ring satisfying no other hypotheses; in some cases he seems to take pride in the fact that he makes no assumption concerning the existence of an identity element. More than any other person, Shinziro Mori is responsible for having worked out the fine details of problems suggested by Refs. 19 and 31.

ACKNOWLEDGMENTS

The author acknowledges with thanks the interest and assistance of many other people in the preparation of this article. In particular, William Heinzer, Takasi Kusano, Joe Mott, Masayoshi Nagata, Jürgen Neukirch, Paulo Ribenboim, and Martha Smith were helpful in the author's efforts.

REFERENCES

Akizuki, Y. (1935). Teilerkettensatz und Vielfachenkettensatz, *Proc. Phys.-Math. Soc. Japan 17*, 337-345.

Birkhoff, G. (1976). The rise of modern algebra to 1936, in *Men and Institutions in American Mathematics*, Texas Tech University.

Chevalley, C. (1943). On the theory of local rings. *Ann. Math. 44*, 690-708.

Cohen, I. S. (1950). Commutative rings with restricted minimum condition, *Duke Math. J. 17*, 27-42.

Dick, A. (1970). *Emmy Noether*, Birkhäuser, Basel.

Dieudonné, J. (1970). The work of Nicholas Bourbaki, *Amer. Math. Monthly 77*, 134-145.

Dirichlet, P. G. L. (1894). *Vorlesungen über Zahlentheorie* (presented and supplemented by Richard Dedekind), Vierte Auflage, Braunschweig.

Fraenkel, A. (1914). Über die Teiler der Null und die Zerlegung von Ringen, *J. reine angew. Math. 145*, 139-176.

Gilbert, J. R., and H. S. Butts (1968). Rings satisfying the three Noether axioms, *J. Sci. Hiroshima Univ. Ser. A-I Math. 32*, 211-224.

Hilbert, D. (1890). Über die Theorie der algebraischen Formen, *Math. Ann. 36*, 471-534.

———(1893). Über die vollen Invariantensysteme, *Math. Ann. 42*, 313-373.

———(1894-1895). Die Theorie der algebraischen Zahlkörper, *Jber. Deutsch. Math.-Verein. 4*, 175-546.

Kaplansky, I. (1973). *Commutative Rings*, Lecture Notes in Mathematics 311, Springer-Verlag, New York.

Kimberling, C. (1972). Emmy Noether, *Amer. Math. Monthly* 79, 136-149.
Kronecker, L. (1882). Grundzüge einer arithmetischen Theorie der algebraischen Grössen, *J. reine angew. Math.* 92, 1-123.
Krull, W. (1922). Algebraische Theorie der Ringe I, *Math. Ann.* 88, 80-122.
─────(1923). Ein neuer Beweis für Hauptsätze der allgemeinen Idealtheorie, *Math. Ann.* 90, 55-64.
─────(1948). *Idealtheorie,* Ergebnisse der Mathematik und ihrer Grenzgebiete, Vierter Band, Chelsea, New York.
─────(1925). Über verallgemeinerte endliche abelschen Gruppen, *Math. Z.* 23, 161-196.
Kubo, K. (1940). Über die Noetherschen fünf Axiome in kommutativen Ringen, *J. Sci. Hiroshima Univ. Ser. A.* 10, 77-84.
Lasker, E. (1905). Zur Theorie der Moduln und Ideale, *Math. Ann.* 60, 20-116.
Macaulay, F. S. (1913). On the resolution of a given modular system into primary systems including some properties of Hilbert numbers, *Math. Ann.* 74, 66-121.
Matusita, K. (1944). Über ein bewertungstheoretisches Axiomensystem für die Dedekind-Noethersche Idealtheorie, *Japan J. Math.* 19, 97-110.
Schmidt, F. K. (1936). Über die Erhaltung der Kettensätze der Idealtheorie bei beliebigen endlichen Körpererweiterungen, *Math. Z.* 41, 443-450.
Sono, M. (1917). On congruences, *Mem. Coll. Sci. Univ. Kyoto 2*, 203-226.
─────(1918). On congrences II, *Mem. Coll. Sci. Univ. Kyoto 3*, 113-149.
Steinitz, E. (1910). Algebraische Theorie der Körper. *J. reine angew. Math.* 137, 167-308.
van der Waerden, B. L. (1931). *Moderne Algebra,* Vol. II. Grundl. Math. Wiss. in Einzeldarst., Band 34, Springer-Verlag, Berlin.
Weyl, H. (1935). Emmy Noether, *Scripta Math.* 3, 201-220.

9

Representation Theory

T. Y. Lam[1]

> *She originated above all a new and epoch-making style of thinking in algebra.*

In these words, Hermann Weyl (1935) described Emmy Noether's legacy in mathematics. A mathematical historian cognizant of the development of modern algebra in the last fifty years will probably find little reason to disagree.

A cogent illustration for Weyl's observation is, of course, Noether's revolutionary idea of working "abstractly" (or "axiomatically") with rings and their ideals. This point of view, first advanced by Noether and later further developed by W. Krull, essentially created the subject of what we now call commutative algebra. For a detailed discussion of Noether's work in this area, we refer the reader to Gilmer's article in this volume. In the present article, we shall focus instead on the importance of Noether's ideal-theoretic viewpoint when applied to *noncommutative* rings, especially group rings and hypercomplex systems. Noether's penetrating ideas in this area have wrought fundamental changes in the theory of representations of finite groups. We shall give an account of her contributions to this subject, and make an attempt to put them somewhat in a historical perspective.

To begin our story, we go back about 55 years into history. The time was the mid-1920s; the place was Göttingen, Germany, then uncontested mathematical capital of the world. The legendary Felix Klein had just died; the equally legendary David Hilbert was in his sixties, interested much more in logic and mathematical physics than in algebra. It had fallen to Emmy Noether to become the de facto leader of the traditionally strong algebra school at Göttingen. Male chauvinism (borrowing a modern cliché) had sufficiently abated so that Noether, with the title of Nicht-beamteter Ausserordentlicher Professor, could now announce her lectures in her own name. With the award of a Lehrauftrag in algebra, she was even able to derive a modest salary, about which the good-natured Emmy seemed to have been delighted. By then, she had published her fundamental papers on "allgemeine Idealtheorie" in commutative rings [19, 24, 31].* Now in her forties but just beginning to reach the peak of her great creative powers, Noether was to waste no time. Before her contemporaries could even grasp the true depth of her abstract ideal theory in the commutative case, Noether's mathematical interest already began to shift in another direction—the ideal theory of hypercomplex systems and its applications to group representations.

Ideals in (and modules over) noncommutative rings are hardly any less natural than their counterparts in the commutative case. Noether realized this and saw that it would be profitable to pursue a noncommutative ideal theory as a means of studying arithmetic questions. In the specific case of group rings over finite groups, such an ideal theory would encompass and elucidate the classical representation theory of G. Frobenius, giving it (in Noether's own words) a "purely arithmetical foundation." This circle of ideas began to emerge in Noether's Göttingen lectures from 1924. The year after, Noether published a short note [28] in the *Jahresbericht* of the German Mathematical Society, announcing her ideas on "group characters and hypercomplex systems." The complete account of this work, however, appeared only four years later. As was usual with Noether, priority was never uppermost in her mind; rather, [paraphrasing Weyl (1935)], her style of research was to expound her ideas to her followers in an unfinished form, lecturing one semester after another on her research topic, until the subject matter reached a pure and unified state. This was particularly true of her approach to representation theory. Her ideas had been fermenting in her lectures for a few years; then, in the winter semester of 1927-1928, "in full clarity and generality" (van der Waerden 1935), she presented to her Göttingen audience her work on hypercomplex numbers and representation theory. Her lectures, redacted with the collaboration of B. L. van der Waerden, appeared in print as Ref. 35^2 (cf. also Ref. 34). Slightly later (1931), van der Waerden completed his *Moderne Algebra*, based on lectures of Artin and Noether. In Chapters 16 and 17 of this text, in particular, there is a full and polished account of hypercomplex systems and their representations, using Noether's ideal/module-theoretic viewpoint in Ref. 35. The timelessness of these two

---

*Numbered references are to Noether's papers listed at the end of this book.

chapters is perhaps the best testament to Noether's profound influence on the subject: many a modern student in algebra still learns representation theory à la Noether and van der Waerden—via modules, ideals, direct decomposition, semisimplicity, DCC rings, etc.

Before we survey Ref. 35 in more detail, it is perhaps worthwhile to say a few words about group representations in physics. Almost in exact parallel to Noether's work on finite group representations, the theoretical physicists at Göttingen (Born, Heisenberg) were making important discoveries in quantum mechanics. These discoveries, in turn, prompted J. von Neumann, E. Wigner, and H. Weyl to develop a representation theory for continuous groups (in infinite as well as finite dimensional Hilbert spaces), study its relations to harmonic analysis, and derive its applications to quantum theory.[3] In the same winter semester of 1927-1928 when Noether lectured on hypercomplex systems at Göttingen, Hermann Weyl was giving lectures at the Eidgenössische Technische Hochschule in Zürich on group theory and quantum mechanics, from which his famous book developed. The winter semester one year before, Weyl was Visiting Professor at Göttingen and gave lectures on representations of continuous groups, with Noether in the audience. There must have been considerable intellectual interaction between the two, as Weyl fondly recalled (1935, p. 208) the many discussions with Noether when they walked home after the lectures, in the "cold, dirty, rain-wet streets" of Göttingen. Though Noether appeared to have played no active role in the application of representation theory to physics, the parallel course of this development and Noether's own "arithmetic" formulation of representation theory witnessed nevertheless a remarkable synchrony of ideas in physics and mathematics.

We now return to representations of finite groups. To appreciate the significance of Noether's work in this area, it would be useful to recall the origin of representation theory, and the historical forces which brought this theory into existence.[4] In doing so, one goes back inevitably to the notion of characters of finite abelian groups. The earliest use of characters is probably the Legendre symbol and its generalization, the Jacobi symbol. In a different vein, Gauss used characters on groups of binary quadratic forms to distinguish among different genera of forms. Dirichlet, in developing his analytic theory of numbers, made substantial use of characters, and invented the functional notation for them. Prompted by this, Dedekind defined a general character on an (abelian) group to be a multiplicative map from the group into the nonzero complex numbers. This definition appeared in the third edition (1878) of *Vorlesungen über Zahlentheorie* by Dirichlet and Dedekind, as well as three years later in the work of Weber.

In 1880, motivated by number theory, Dedekind was led to an interesting determinant which one can associate to any finite group G: let $\{x_g\}$ be independent indeterminates indexed by $g \in G$, and consider the matrix whose (g,h)-entry is $x_{gh^{-1}}$. Its determinant $D = D\{x_g : g \in G\}$ is called the *group determinant* of G. Dedekind made the ingenious discovery that in case G is *abelian*, $D\{x_g\}$ factors completely into linear forms over the complex numbers by using the characters of G, namely,

$$D = \prod_{\chi \in \hat{G}} \left(\sum_{g \in G} \chi(g) x_g \right) \qquad \hat{G} = \text{character group of } G \qquad (*)$$

The proof is not difficult. Indeed, for a fixed $\chi \in \hat{G}$, multiply the gth row of the determinant by $\chi(g)$ and add up all rows. On the hth column, we get

$$\sum_{g \in G} \chi(g) x_{gh^{-1}} = \left[\sum_{g' \in G} \chi(g') x_{g'} \right] \chi(h)$$

Thus, D is divisible by $\sum_{g \in G} \chi(g) x_g$ for each character χ. Since there are $|G|$ different characters and these give rise to different linear forms, we get (*).

But now a puzzling question arises: what if G is not abelian? The above proof will nevertheless give $|G/[G,G]|$ different linear factors but these will no longer "exhaust" the group determinant. Dedekind perhaps sensed that it would be important to answer the above question, because the "missing" factors of D in the nonabelian group case surely must contain vital information about the group. Years later, between March and April 1896, Dedekind wrote two letters to G. Frobenius and explained this problem, hoping to arouse Frobenius' interest in it. These letters, as it turned out, became the catalyst for the character theory of arbitrary finite groups.

Frobenius was a great expert on determinant theory. Also, by happy coincidence, in one of his earlier works on theta functions, he had dealt with a somewhat similar determinant. After Frobenius received the Dedekind letters, a cordial correspondence ensued, the remarkable result of which was that Frobenius completely solved the group determinant problem by creating a beautiful and far-reaching theory of "higher dimensional characters" of finite groups ! Frobenius (1896a,b) showed that the different prime factors of the group determinant D correspond one-to-one to his new "primitive" (or irreducible) characters χ; each such factor has degree equal to $\chi(1)$ (the degree of the character), and occurs in D exactly to multiplicity $\chi(1)$. These degrees always divide $|G|$ and satisfy the equation $|G| = \Sigma \chi(1)^2$.

A year later, Frobenius (1897) also discovered the more familiar description of characters as we know them today, using the trace of matrix representations.

Few would deny that Frobenius' analysis of the group determinant (and the concomitant discovery of character theory) is a mathematical tour de force; however, a modern student in representation theory will probably find this approach to character theory rather unnatural, if not bizarre. One question seems inevitable: if Dedekind had not written to Frobenius about the group determinant, would character theory still have been discovered, and if so, when?

An eloquent answer to this question is to be found in Hawkins' article (1971-1972b). In this work, Hawkins developed the theme that "the creation of group representation theory was linked with a broad spectrum of late 19th century mathematical thought." In Hawkins' analysis, Frobenius' ingenious work was but one of several lines of investigation which had led to group representation theory. The other, almost parallel, lines of investigation are: (1) the structure theory of hypercomplex systems by Molien and Cartan, (2) the Lie-theoretic methods in studying group algebras by Burnside, and (3) the discovery of complete reducibility of matrix representations by Maschke.

We shall not go into (2) and (3) here, but we must discuss Molien's work in (1) because it was perhaps the most important predecessor to Noether's paper Ref. 35. The Estonian mathematician Theodor Molien did not have a very successful professional career; his name is hardly mentioned in books on representation theory. But anyone patient enough to analyze his difficult papers will not fail to recognize the high degree of originality in his ideas. In his thesis (1893) Molien dealt with the structure theory of hypercomplex systems. Working over the complex numbers as the base field, he arrived at the notion of simple and semisimple algebras, and obtained the decomposition of semisimple algebras into matrix algebras[5]–15 years before Wedderburn proved the same result over arbitrary fields. In two subsequent papers (1897a,b), applying his structure theory to complex group algebras, Molien arrived at many of the basic theorems of representation theory independently of Frobenius.

Frobenius himself fully recognized the importance of Molien's work, and did not hesitate to praise it. Upon learning that Molien was still only a Privatdozent in Dorpat, Frobenius even wrote to the influential Dedekind, hoping that he might help to further Molien's career. Nevertheless, Molien's work remained in relative obscurity; the significance of his method and its potential for generalization went essentially unnoticed.

In a way, Noether's paper Ref. 35 changed that. Molien's work had a definite influence on Noether's: the idea of developing representation theory as a part of the structure theory of hypercomplex systems had been a central theme in Molien's papers. But in the 1920s, Noether had much more to go upon. On the one hand, the general Wedderburn theorem over arbitrary fields was available; on the other hand, ideal theory in noncommutative rings had become much better understood. Instead of working clumsily with ideals as Molien did in the 1890s, Noether possessed the insight to recognize the exact relationship between ideals and representations, and to use this relationship in the light of the Wedderburn theory to recast the foundations of group representation theory.

Central to Noether's new approach are the basic ideas of modules, submodules, direct sums, homomorphisms, etc. She pioneered the use of a representation space: if G is a group and k a field, Noether thought of a k-representation of G as a k-vector space V with G acting as a group of k-linear transformations. To go one step further, Noether viewed V as a module over the group algebra kG, and called it a *representation module*. Two representation modules will give equivalent representations of G iff they are isomorphic as kG-modules. Irreducible representations correspond to simple kG-modules, and decomposition of representations corresponds to direct sum decomposition of kG-modules. Left ideals $\mathfrak{A} \subseteq kG$ are special cases of kG-modules and so they afford k-representations of G; the representation is irreducible iff $\mathfrak{A}$ is minimal as a left ideal. These new viewpoints create a very effective conceptual framework in which to study the representation theory of groups.

In this module-theoretic approach to representation theory, clearly the group algebra kG plays a crucial role—and this is where hypercomplex systems come in. If one develops a good theory of modules over hypercomplex systems, facts in group representations will follow as consequences. Here, we see that Noether picked up exactly where Molien left off. But in Ref. 35, Noether developed an even more general theory—she went beyond finite dimensional hypercomplex systems, and worked in the larger "category" of the so-called DCC rings.

Shortly after Noether's effective use of the maximum condition (ACC) on ideals, Artin (1927) had pointed out the role of the "dual" minimum condition (DCC). In Ref. 35, acknowledging the work of Artin (1927) and Suschkewitsch (1928), Noether developed the Wedderburn theory for DCC rings without a radical. She showed that

(a) Such rings are precisely the semisimple (or completely reducible) rings (every left ideal is a direct summand).

Representation Theory

(b) These rings are (uniquely) a finite direct sum of indecomposable two-sided ideals, each of which is a simple ring with DCC.

(c) Each simple ring A with DCC is isomorphic to a matrix algebra over a division ring D.

The last part (c), in the case of finite dimensional algebras, is, of course, the central pillar of the classical Wedderburn theory. In Noether's module-theoretic framework, (c) acquires a new and especially natural interpretation. Noether showed that the ring A in (c) has, up to isomorphism, a *unique* left simple module M, and that A is a direct sum of (minimal) left ideals isomorphic to M. As for the division ring D in (c), Wedderburn obtained it in a rather noninvariant way as a subring of A, but Noether arrived at D very naturally as the A-endomorphism ring of M (provided that we write endomorphisms on the right of elements $m \in M$). The fact that D is a division ring is the well-known Schur's lemma. Viewing M as a right D-vector space, its D-linear endomorphisms are given precisely by left multiplications by elements of A, so A is isomorphic to a matrix algebra over D.[6]

If A is semisimple instead of simple, we decompose A into $A_1 \oplus \cdots \oplus A_r$ as in (b). Viewing A_i as a quotient ring of A, the simple A_i-module M_i is also a simple A-module; Noether showed that, up to isomorphisms, *all* simple A-modules arise in this way. If A is only DCC but not semisimple, its simple modules are the same as those for the semisimple quotient ring $A/\mathrm{rad}(A)$, and are therefore completely accounted for by the above theory.

When applied to group algebras kG (k a field, G a finite group), Noether's reformulation of the Wedderburn theory above gives a particularly clear picture of the representation theory of G. If the characteristic of k is prime to the order of G, kG is semisimple by Maschke's theorem, so the "left regular module" kG decomposes into $n_1 M_1 \oplus \cdots \oplus n_r M_r$ where $M_1, \ldots, M_r$ are the different irreducible kG-representations. Comparing degrees, we have an equation $|G| = n_1^2 d_1 + \cdots + n_r^2 d_r$ where d_i are the k-dimensions of the kG-endomorphism rings of M_i. In the special case when k is algebraically closed, these division rings all coincide with k, so we get back Frobenius' equation $|G| = n_1^2 + \cdots + n_r^2$, where the n_i are now the degrees of the irreducible representations; r, the number of irreducible representations, equals the k-dimension of the center of kG, which is the number of conjugacy classes of G. Using the decomposition $kG \cong n_1 M_1 \oplus \cdots \oplus n_r M_r$, it is also possible to give a quick derivation and a natural explanation for Frobenius' factorization of Dedekind's group determinant mentioned earlier.

Thus, in a very succinct way, Noether's theory captures the basic facts about finite group representations. More importantly, while the classical approach to representation theory is valid only over algebraically closed fields of characteristic zero, Noether's approach remains meaningful for non-algebraically closed fields of arbitrary characteristic. The importance of considering representations over fields whose characteristic may divide the order of the group became apparent later in Brauer's theory of modular representations, though we shall not go into this topic here. In the modular case, Noether's viewpoint also suggests naturally the study of *indecomposable* kG-modules as a generalization of irreducible ones. This study, however, is vastly more complicated than classical representation theory, and remains very far from being complete even today. More generally, it is of interest to study the classification of indecomposable modules over nonsemisimple finite dimensional algebras (and DCC rings). This has been a very active field of research in modern algebra; for some of the advances made in recent years, see the relevant articles in Dlab and Gabriel (1975), and the bibliographies contained therein.

In another direction, Noether's idea of representation modules also planted the seed for modern integral representation theory. For any (commutative) ring k, kG-modules make perfect sense and we can regard them as giving k-representations of G. These modules need not be k-free, so the representations which they afford may no longer be described by matrices. Nevertheless, we can study them in the general framework created by Noether—using submodules, exact sequences, homomorphisms, endomorphisms, etc. In dealing with integral representations, a useful modern device is to form "Grothendieck groups" out of various classes of kG-modules; such a device is by and large compatible with Noether's view of representation theory. Today, integral representation theory is an important subject for research. Through interactions with algebraic K-theory, it has found deep connections to many areas in topology, for instance, the study of homotopy types and surgery of manifolds. The diversity of such applications was perhaps totally unanticipated by Noether.

[For a survey of integral representation theory, see Curtis and Reiner (1962, Ch. 11) and Reiner (1970, 1980); the former is a standard work which especially espouses Noether's module-theoretic viewpoint. For the interplay between integral representations and algebraic K-theory, see, e.g., Swan and Evans (1970).]

Returning now to representations over fields, we shall discuss another interesting contribution of Noether, in the joint paper Ref. 33 with Richard Brauer, then a young Privatdozent at Königsberg. In this work, Brauer and Noether clarified the notion of splitting fields for irreducible representations, relating them to splitting fields of simple algebras and division algebras. This work is directly relevant to the important notion of the Schur index, introduced by Schur (1906) some 20 years earlier. To discuss the Brauer-Noether paper, let us first recall the main ideas behind the Schur index.

In the first phase of group representation theory (as in the work of Frobenius, Molien, and Burnside), mathematicians were primarily concerned about representations over the complex field; "irreducibility" of a representation always means irreducibility over the complex numbers. In the first decade of the 20th century, investigators began to pay closer attention to the fields over which a representation is defined. The first systematic effort at studying the behavior of representations under base field change was undertaken by Schur (1906). In this classical work, Schur initiated a beautiful theory of "indices" for group characters χ given by absolutely irreducible representations. The Schur index of χ relative to a base field k contains very rich and valuable arithmetic infomation about χ itself as well as about the field k.

The definition of the Schur index runs as follows. Let G be a finite group and k be a field, say of characteristic zero, with algebraic closure $\bar{k}$. Let V be an irreducible kG-module, and $\bar{V}$ be its scalar extension to $\bar{k}G$. Then, according to Schur's theory, the irreducible characters χ occurring in $\bar{V}$ are "algebraically conjugate" over k, each occurring with the same multiplicity. This common multiplicity is called the *Schur index* of χ relative to k, denoted by $m_k(\chi)$.

The Schur index is intimately tied to the study of the splitting fields for V: Recall that an (algebraic) extension $K \supseteq k$ is a *splitting field* for V if the scalar extension of V to K splits up into absolutely irreducible components (or, equivalently, if each χ in the above paragraph is afforded by a representation of G over K). The relationship between $m_k(\chi)$ and the splitting fields for V is as follows: Let $k(\chi)$ be the field obtained from k by adjoining all values of χ. Then, for any splitting field $K \supseteq k$ for V, $[K : k(\chi)]$ is a multiple of $m_k(\chi)$; moreover, there exists at least one splitting field K_0 for V with $[K_0 : k(\chi)] = m_k(\chi)$.

Schur's paper is a highly original piece of work; it dealt with a topic which was completely new to his contemporaries; yet the results were remarkably penetrating and complete. This important paper of Schur was to inspire much subsequent work on splitting fields, real representations, etc.

Schur's proof of the various theorems concerning the index was based largely on manipulations of the group determinant, in the spirit of Frobenius. But in Ref. 33, Brauer and Noether recast Schur's theory in the language of splitting fields of central (or "normal") simple algebras. Keeping our earlier notations, let $D = \text{End}_{kG} V$; by Schur's lemma, this is a division algebra over k. It can be shown that the center of D is precisely the field $k(\chi)$; thus, D is a central $k(\chi)$-division algebra. Brauer and Noether noted that the splitting fields for V are precisely the splitting fields for this algebra, i.e., fields $K \supseteq k(\chi)$ such that $K \otimes_{k(\chi)} D$ is a matrix algebra over K. Thus, remarkably, the notion of splitting an irreducible representation and the notion of splitting a central simple algebra go hand-in-hand. This viewpoint lends itself nicely to elucidate Schur's theorems on the index, as follows. One knows that the "smallest" splitting fields K_0 for D are precisely the maximal subfields of D; their $k(\chi)$-dimensions satisfy the equation $[\dim_{k(\chi)} K_0]^2 = \dim_{k(\chi)} D$. These K_0 are smallest in the strong sense that any splitting field K of D has $\dim_{k(\chi)} K$ divisible by $\dim_{k(\chi)} K_0$. Via this description of K_0, we see that $\dim_{k(\chi)} K_0$ is precisely the Schur index $m_k(\chi)$. It equals $[\dim_{k(\chi)} D]^{1/2}$ which, therefore, is called the Schur index of the $k(\chi)$-division algebra D. In this way, we achieve a harmonious blending of the theory of group representations and the theory of central simple algebras.

A splitting field K for D is called *minimal* if it has no proper subfields which are splitting fields. Brauer and Noether showed that, though K may be minimal as a splitting field, its dimension over $k(\chi)$ may not be the Schur index $m_k(\chi)$; in fact, it can be an arbitrarily large multiple of $m_k(\chi)$! By working with central simple algebras, such an example is not difficult to construct. For instance, let $k = \mathbf{Q}$ (the rational field), and G be the quaternion group of 8 elements. The group algebra $\mathbf{Q}G$ splits into five simple components: $\mathbf{Q} \oplus \mathbf{Q} \oplus \mathbf{Q} \oplus \mathbf{Q} \oplus D$ where D is Hamilton's quaternion algebra, viewed over the rationals. This last component affords an irreducible $\mathbf{Q}G$-representation $V = D$, which, when viewed over $\mathbf{C}$, gives a character 2χ where χ is the unique irreducible character of dimension 2. This shows that $m_\mathbf{Q}(\chi) = 2$ and $\mathbf{Q}(\chi) = \mathbf{Q}$. What are the splitting fields for V? By what we have said before, these are just the splitting fields for D; since D has the norm form $x^2 + y^2 + z^2 + w^2$, the splitting fields for D are easily seen to be extensions of $\mathbf{Q}$ in which -1 is a sum of two squares. (Clearly these fields must have *even* degree over $\mathbf{Q}$ since odd extensions of $\mathbf{Q}$ may be imbedded into the real numbers!) Using a bit of number theory, it is possible to show that there exist $K \supseteq \mathbf{Q}$ of arbitrarily large degree (in fact of any even degree) minimal with respect to the property that -1 is a sum of two squares in K. Thus, there is no bound to the size of minimal splitting fields of the irreducible $\mathbf{Q}G$-representation module V.

The Brauer-Noether paper seemed to have marked the beginning of Noether's interest in central simple algebras; she returned to this topic again several times in her later work. In Ref. 41, Sec. 7-9, for example, she treated splitting fields of algebras in greater detail, and in Ref. 39, Brauer, Hasse, and Noether completed the program of classifying central division algebras over algebraic number fields.

NOTES

1. Supported in part by NSF.
2. This paper appeared in *Math. Zeit.* instead of in *Math. Annalen* as originally announced in Ref. 28.
3. For more details on this, we refer the reader to Mackey's excellent survey (1980), especially Section 16.
4. For more details and documentations on this, we refer the reader to the three erudite articles by T. Hawkins (1971-1972a,b; 1974) which are hard to improve upon. We shall present here only a summary of Hawkins' discussions.
5. Similar results were obtained slightly later (but independently) by E. Cartan. The work of Molien and Cartan was, in part, motivated by Killing's classification of simple Lie algebras, obtained several years earlier.
6. This formulation of Noether was further generalized into a "Density Theorem" later by Jacobson and Chevalley; see, e.g. Artin's article (1950) for a clear exposition.

REFERENCES

Artin, E. (1927). Zur Theorie der hyperkomplexen Zahlen, *Abh. Math. Sem. Univ. Hamburg 5*, 307-316.

―――――― (1950). The influence of J. H. M. Wedderburn on the development of modern algebra, *Bull. Amer. Math. Soc. 56*, 65-72.

Curtis, C.W., and I. Reiner (1962). *Representation Theory of Finite Groups and Associative Algebras*, Interscience, John Wiley, 1962.

Dlab, V., and P. Gabriel, eds. (1975). *Representations of Algebras, Proc. of the International Conference, Ottawa, 1974*, Springer Lecture Notes in Mathematics, Vol. 488, Springer-Verlag, Berlin-Heidelberg-New York.

Frobenius, G. (1896a). Über Gruppencharaktere, *Sitzungsber. Akad. d. Wiss. Berlin*, 985-1021.

―――――― (1896b). Über die Primfactoren der Gruppendeterminante, *Sitzungsber. Akad. d. Wiss. Berlin*, 1343-1382.

―――――― (1897). Über die Darstellung der endlicher Gruppen durch lineare Substitutionen, *Sitzungsber. Akad. d. Wiss. Berlin*, 944-1015.

Hawkins, T. (1971-1972a). The origins of the theory of group characters, *Archive for History of Exact Sciences 7*, 142-170.

―――――― (1971-1972b). Hypercomplex numbers, Lie groups and the creation of group representation theory, *Archive for History of Exact Sciences 8*, 243-287.

―――――― (1974). New light on Frobenius' creation of the theory of group characters, *Archive for History of Exact Sciences 12*, 217-243.

Mackey, G. (1980). Harmonic analysis as the exploitation of symmetry—a historical survey, *Bull. Amer. Math. Soc. 3*(new series), 543-698.

Molien, T. (1893). Über Systeme höherer complexer Zahlen, *Math. Ann. 41*, 83-165. [Berichtigung: *Math. Ann. 42*(1893), 308-312.]

——— (1897a). Eine Bemerkung zur Theorie der homogenen Substitutionsgruppen, *S'ber. Naturforscher-Ges. Univ. Jurjeff* (Dorpat) *11*, 259-274.

——— (1897b). Über die Anzahl der Variabeln einer irreductibelen Substitutionsgruppe, *S'ber. Naturforscher-Ges. Univ. Jurjeff* (Dorpat) *11*, 277-288.

Reiner, I. (1970). A survey of integral representation theory, *Bull. Amer. Math. Soc. 76*, 159-227.

——— (1980). An overview of integral representation theory, in *Ring Theory and Algebra III* (ed. B. McDonald), 269-300, Lecture Notes in Pure and Applied Mathematics, Vol. 55, Marcel Dekker, New York.

Schur, I. (1906). Arithmetische Untersuchungen über endliche Gruppen, *Sitzungsber. Akad. d. Wiss. Berlin,* 164-184.

Suschkewitsch, A. (1928). Über die endlichen Gruppen ohne das Gesetz der eindeutigen Umkehrbarkeit, *Math. Ann. 99*, 30-50.

Swan, R., and E. G. Evans (1970). *K-Theory of Finite Groups and Orders,* Springer Lecture Notes in Mathematics, Vol. 149, Springer-Verlag, Berlin-Heidelberg-New York.

van der Waerden, B. L. (1935). Nachruf auf Emmy Noether, *Math. Ann. 111*, 469-476.

Weyl, H. (1935). Emmy Noether, *Scripta Math. 3*, 201-220.

10

Algebraic Number Theory

A. Fröhlich

A great deal of Emmy Noether's work involves algebraic number theory. It enters as a tool into the investigation of splitting fields for group representations (cf. Ref. 33)*, and it provides the motivation of her work in Galois theory (cf. Ref. 11) and of some of her work in commutative ring theory (e.g., Refs. 31, 32). The theory of crossed products and central simple algebras has been a main tool in class field theory and it is in this area that her famous joint paper with Brauer and Hasse lies (cf. Ref. 39). Moreover, Noether also made a contribution to the arithmetic of such crossed products (cf. Ref. 43), a sort of noncommutative number theory. Many of these aspects are dealt with elsewhere in this volume and we shall concentrate on her work (i) on Galois module structure (cf. Ref. 38) and (ii) on what we now would call the cohomology of number fields (cf. Ref. 42). Her report (cf. Ref. 40) to the Zürich International Congress gives a survey of both these topics.

GALOIS MODULE STRUCTURE

In this area Noether was well ahead of her time. She seems to have been almost alone in realizing the significance and genuine interest of the Galois module structure of rings of algebraic integers or of their local counterparts, a topic which others, both at the time and for many years to come, apparently

---

*Numbered references are to Noether's papers listed at the end of this book.

considered devoid of any deeper content. Noether was one of the first (following some results of Hilbert and of Speiser) to see the connection between module properties and ramifications, which is now the subject of an extensive theory, and she was the only one to guess the interpretation of various arithmetic character invariants—in her case of the Artin conductors—in terms of Galois module structure. It is this tentative anticipation of developments to come 40 years later, rather than just the theorem given by her at the time, which secures her an important place in the history of the subject.

To explain the contents of Ref. 38 we first recall the basic definitions in ramification theory and we do this in a local context. Let k be a local field, to be definite let us say an extension of finite degree of a rational p-adic field $\mathbf{Q}_p$, and let K be an extension of finite degree of k. Denote by $\mathfrak{o}_K$ and $\mathfrak{o}_k$ the respective rings of integers (valuation rings) in K and k, and by $\mathfrak{p}_K$ and $\mathfrak{p}_k$ their maximal ideals. Then $\mathfrak{p}_k \mathfrak{o}_K = \mathfrak{p}_K^e$ for some $e \leqslant 1$. If $e = 1$, K/k is nonramified; if $p \nmid e$ then we say that K/k is tame—in Noether's language there is no "higher ramification." If K/k is Galois then we get criteria for these properties in tems of the Galois group Γ. We define

$$\Gamma_0 = \{\gamma \in \Gamma: X^\gamma - X \in \mathfrak{p}_K \text{ for all } X \in \mathfrak{o}_K\}$$
$$\Gamma_1 = \{\gamma \in \Gamma: X^\gamma - X \in \mathfrak{p}_K^2 \text{ for all } X \in \mathfrak{o}_K\} \quad (10.1)$$

Then Γ_0 and Γ_1 are normal subgroups of Γ, and Γ_1 is the unique p-Sylow group of Γ_0. The criteria are (1) K/k is nonramified if and only if $\Gamma_0 = 1$, (ii) K/k is tame if and only if $\Gamma_1 = 1$.

In the global case, e.g., for a Galois extension E/F of algebraic number fields, one can make the analogous definition directly, or one goes over to local extensions $E_\mathfrak{P}/F_\mathfrak{p}$ where $\mathfrak{p}$ is a prime ideal in F, $\mathfrak{P}$ one in E above $\mathfrak{p}$. The local Galois group is then embedded in the global one, and is—together with its subgroups Γ_0, Γ_1—determined to within conjugacy by $\mathfrak{p}$ alone.

To understand the background of the problem Noether was looking at, recall that if K/k is any Galois extension of fields of finite degree, with Galois group Γ, then K has a *normal basis* $\{a^\gamma\}_{\gamma \in \Gamma}$ over k, consisting of the conjugates of a fixed element a, or in other words, K is a free module over the group ring $k\Gamma$ on a generator a. If now $\mathfrak{o}_k$ is a "ring of integers" in k and $\mathfrak{o}_K$ its integral closure in K, one can ask whether a *normal integral basis*, i.e., a basis $\{a^\gamma\}_{\gamma \in \Gamma}$ of $\mathfrak{o}_K$ over $\mathfrak{o}_k$ exists. The first (global) result in the literature is the theorem of Hilbert-Speiser that if $k = \mathbf{Q}$, the rational field, with $\mathfrak{o}_k = \mathbf{Z}$, and if $K \subset \mathbf{Q}(e^{2\pi i/m})$, m square-free, then indeed a normal integral basis

Algebraic Number Theory

exists. These fields K are precisely those extension fields of $\mathbf{Q}$ which have abelian Galois groups and which are tame (at all primes) (cf. Hilbert 1897, Theorem 132). Hilbert actually made the stronger assumption that the discriminant of K is prime to the degree. The more general result is ascribed to Speiser.

Now we return again to the local fields as considered above. By a result of Speiser (1916) the existence of a normal integral basis of $\mathfrak{o}_K$ over $\mathfrak{o}_k$ implies that K/k is tame. The theorem of Noether asserts the converse; in other words, if K/k is tame then $\mathfrak{o}_K$ has a normal integral basis over $\mathfrak{o}_k$. Noether's proof actually contains a gap—for a recent proof see Fröhlich (1967, page 22). Richard Brauer also mentioned to me once in conversation that he had at the time supplied a correct proof, but to my knowledge this has not been published.

The theorems of Speiser and Noether can be restated, in either local or global terms, as establishing an equivalence between tameness of ramification of K/k on the one hand and the property of $\mathfrak{o}_K$ being projective over the group ring $\mathfrak{o}_k\Gamma$ on the other. They thus appear as forerunners of more recent results, which provide us with invariants measuring simultaneously the deviation of ramifications in K/k from tameness and of the Galois module $\mathfrak{o}_K$ from projectivity [cf. Fröhlich (1976)]. In a related but slightly different direction, the determination of the local module structure in the tame case, given by Noether's theorem, has been a precondition for the rapid developments of global Galois module structure theory in the last decade. In the intervening years this topic was largely ignored; one exception is a paper by M. Newman and Olga Taussky (1956); another is Leopoldt's paper (1959).

The comments which in Noether's paper follow the basic theorem form its most interesting part. Noether observes that for a tame extension K/k, the relative discriminant $\mathfrak{d}(K/k)$ of K/k is now the square of the group determinant (which has played a fundamental role in classical representation theory)—of course, with the indeterminates replaced by the conjugates a^γ forming a normal integral basis. But the group determinant can be decomposed in terms of the irreducible characters χ of the Galois group Γ. From this one now deduces a decomposition of its square, i.e., of the discriminant, in the form

$$\mathfrak{d}(K/k) = \Pi_\chi \mathfrak{c}(K/k, \chi) \qquad (10.2)$$

where the $(K/k, \chi)$ are ideals which are certainly defined in $k(\chi)$, the field of values of χ over k. Noether was thus led—well ahead of her time— to guess at a connection between module structure and character invariants of number fields. [See however also Speiser (1916).]

A few years before Noether's paper, E. Artin had defined his conductors, i.e., ideals $f(K/k, \chi)$ associated with characters of the Galois group of K/k, where K/k now need not be tame. The general definition is in terms of a rather complicated formula involving the sequence of ramification subgroups of Γ, the first two of which we defined in (1). See Artin (1931) or Serre (1967, Section 4.3). As in Artin's paper, but taken locally, we view $f(K/k, \chi)$ as an ideal, i.e., $f(K/k, \chi) = \mathfrak{p}_k^{f(\chi)}$, with $f(\chi)$ the integer defined by Serre (1967). In the tame case however the formula is quite simple. Indeed let V be the vector space, with Γ acting by linear transformations, and let V^{Γ_0} be the subspace of fixed points of Γ_0. The conductor of the character χ of the representation is then given by

$$f(K/k, \chi) = \mathfrak{p}_k^{\dim V - \dim V^{\Gamma_0}} \tag{10.3}$$

Noether raises the question whether in fact the module-theoretic factors $\mathfrak{c}(K/k, \chi)$ coincide with the ramification-theoretic factors $f(K/k, \chi)$. (To give a precise formulation of this question one has to identify complex characters of Γ with p-adic ones.) For about 40 years there was no progress towards any solution of this problem. Now, however, the theory of module conductors and resolvents [see in particular Fröhlich (1976) and A. Nelson's thesis] confirm that indeed $f(K/k, \chi) = (K/k, \chi)$, as Noether had evidently believed.

COHOMOLOGY, OR CENTRAL SIMPLE ALGEBRAS

The main aim in Ref. 42 is a new proof, based on the theory of algebras, and a far-reaching generalization of the principal genus theorem of class field theory, which deals with cyclic extensions and which in turn had its roots in Gauss's principal genus theorem on rational quadratic forms. The basic program for Noether's work on this topic, contained in her Congress lecture [40], is to reformulate results of this type in terms of simple algebras, i.e., of crossed products where the restatement makes sense for arbitrary, rather than just cyclic Galois groups, and then to prove them in this much greater generality. This outlook puts Noether well ahead of her time. It was one of the main successes of the cohomological method in later years to show systematically that part of class field theory really deals with arbitrary Galois extensions rather than just abelian ones [cf. Tate (1967)]. In this context the language of cohomology theory and that of the theory of algebras are interchangeable. Indeed Noether gives several formulations for her theorems, one of them in terms of factor systems, which already point the way to the

Algebraic Number Theory

later use of cohomology. We shall restate her results in cohomological translation, so as to help the modern reader appreciate what she did.

Noether begins by proving, for any Galois extension K/k of finite degree with Galois group Γ, that

I. $H^1(\Gamma, K^*) = 0$, K^* the multiplicative group.

Here k is an arbitrary field. This is usually referred to as Hilbert's Theorem 90, but in fact the latter [cf. Hilbert (1897)] only states that for Γ cyclic of prime order, the kernel of the norm map $N_{K/k}: K^* \to k^*$ coincides with $K^{*1-\gamma}$, γ a generator of Γ. Noether shows of course that her theorem reduces in the cyclic case to Hilbert's.

From now on we consider number fields only. As a step to her main theorem, Noether proves

II.   (Hilfssatz 1) $H^1(\Gamma, I) = 0$, I the group of fractional ideals of K.

Her proof is direct. Given a 1-cocycle c in I we let $\mathfrak{b} = \Sigma_{\sigma \in \Gamma} c(\sigma)$ (sum of ideals). Then indeed $c(\sigma) = \mathfrak{b}^{1-\sigma}$.

From here on Noether's results and proofs are very easily restated in terms of cohomology sequences involving idele groups. Let S be the set of primes of K (including infinite ones) ramified in K/k. Let J' and J'' be the subgroups of the idele groups J of K of ideles which have local components = 1 outside S (resp. in S). Then $J = J' \times J''$. We accordingly have maps

$$\pi_1: H^2(\Gamma, J) \to H^2(\Gamma, J') \qquad \pi_2: H^2(\Gamma, J) \to H^2(\Gamma, J'') \qquad (10.4)$$

which make $H^2(\Gamma, J)$ into a product $H^2(\Gamma, J') \times H^2(\Gamma, J'')$. Next we have an injection

$$\iota: H^2(\Gamma, K^*) \to H^2(\Gamma, J) \qquad (10.5)$$

induced by the embedding $K^* \to J$. The fact that ι is injective, i.e., that a central simple algebra which splits everywhere locally also splits globally, is indeed used by Noether. Next we write, with I denoting the ideal group of $\mathfrak{o}_K$ and P the group of principal ideals,

$$\psi: H^2(\Gamma, K^*) \to H^2(\Gamma, P) \qquad \phi: H^2(\Gamma, P) \to H^2(\Gamma, I) \qquad (10.6)$$

for the maps induced from $K^* \to P$ and $P \to I$. Then Noether states

III. (Hilfssatz 2) $\ker \pi_1 \iota \cap \ker \phi\psi = 0$

Proof: With U the group of unit ideles we have an exact sequence

$$0 \to U \to J \to I \to 0$$

and so by II we see that the bottom row in the commutative diagram

$$\begin{array}{ccccc}
& & 0 & & 0 \\
& & \downarrow & & \downarrow \\
0 & \to & \ker \phi\psi \to H^2(\Gamma, K^*) & \to & H^2(\Gamma, I) \\
& & \downarrow \quad\quad\quad \downarrow & & \| \\
0 & \to & H^2(\Gamma, U) \to H^2(\Gamma, J) & \to & H^2(\Gamma, I)
\end{array}$$

is exact. The top row is exact by definition and the columns are exact as ι is injective. The cohomology of local unit groups outside S is trivial. Thus viewed as subgroups of $H^2(\Gamma, J)$, both $H^2(\Gamma, U)$ and its subgroup $\ker \phi\psi$ lie in $\ker \pi_2$. Thus $\ker \phi\psi \cap \ker \pi_1 \iota$ lies in $\ker \pi_1 \cap \ker \pi_2 = 0$, as we had to show.

Let C be the ideal class group of K^*. From the exact sequence

$$0 \to P \to I \to C \to 0$$

and by II we get an exact sequence

$$0 \to H^1(\Gamma, C) \xrightarrow{\delta} H^2(\Gamma, P) \xrightarrow{\phi} H^2(\Gamma, I) \tag{10.7}$$

Noether's generalization of the principal genus theorem then is

IV.   The composite map

$$H^1(\Gamma, C) \to H^2(\Gamma, P) \to H^2(\Gamma, P)/\psi(\ker \pi_1 \iota)$$

is injective.

Proof: Let $x \in H^1(\Gamma, C)$, $a \in \ker \pi_1 \iota$ and $\delta x = \psi a$. Now by exactness of (10.7), $\phi\psi a = 0$. By III, $a = 0$. Hence $x = 0$.

To get back to the classical theorem, suppose Γ is cyclic, on a generator γ. Then $H^1(\Gamma, C)$ is the kernel of the norm map $N_{K/k}$ on ideal classes modulo $C^{1-\gamma}$, and $H^2(\Gamma, P) = \hat{H}^0(\Gamma, P)$ is the group of principal ideals in k modulo the norms of principal ideals. To compute δ one takes an ideal in a given class and evaluates its norm. Finally $\psi(\ker \pi_1 \iota)$ consists of principal ideals (a) in k, where a is a local norm at all ramified primes.

Thus one does indeed get, in the cyclic case, the principal genus theorem:

V. Let $\mathfrak{a}$ be an ideal in K whose norm $N_{K/k}\,\mathfrak{a}$ is principal, $N_{K/k}\,\mathfrak{a} = (a)$, with $a \in k^*$, and moreover a a norm locally at all ramified primes (including infinite ones). Then the ideal class of $\mathfrak{a}$ is of form $C^{1-\gamma}$.

REFERENCES

Artin, E. (1931). Die gruppentheoretische Struktur der Diskriminanten algebraischer Zahlkörper, *J. für die reine u. angew. Math. 164*, 1-11.
Fröhlich, A. (1967). Local Fields, in *Algebraic Number Theory—Brighton Proceedings*, Academic Press, London.
———(1976). Module conductors and module resolvents, *Proc. London Math. Soc. 32*, 279-322.
Hilbert, D. (1897). *Zahlbericht, Gesammelte Abhandlungen I*, Springer, Berlin, 1932. Also Chelsea, New York, 1965.
Leopoldt, H. W. (1959). Über die Hauptordnung der ganzen Elemente eines Abelschen Zahlkörpers, *J. für die reine u. angew. Math. 20*, 119-149.
Newman, M., and O. Taussky (1956). On a generalization of the normal basis in Abelian algebraic number fields, *Comm. Pure Appl. Math. 9*, 85-91.
Speiser, A. (1916). Gruppendeterminante und Körperdiskriminante, *Math. Ann. 77*, 546-562.
Serre, J-P. (1967). Local class field theory, in *Algebraic Number Theory—Brighton Proceedings*, Academic Press, London.
Tate, J. T. (1967). Global class field theory, in *Algebraic Number Theory—Brighton Proceedings*, Academic Press, London.

IV

NOETHER'S ADDRESS TO THE 1932 INTERNATIONAL CONGRESS OF MATHEMATICIANS

11
Hypercomplex Systems and Their Relations to Commutative Algebra and Number Theory

Emmy Noether

1. During the past few years the theory of hypercomplex systems, of algebras, has made good progress; but it is only very recently that the significance of this theory for commutative problems has become evident. Today I should like to comment on this significance of the noncommutative for the commutative: and indeed, I want to do this particularly in view of two classical problems originating from Gauss, the principal genus theorem and the closely related norm principle. The formulation of these problems has undergone continuous change: with Gauss they appear as the culmination of his theory of quadratic forms; then they play an essential role in the characterization of the relative cyclic and abelian number fields via class field theory; and finally they manifest themselves as theorems on automorphisms and on the decomposition of algebras, and at the same time this last formulation allows the theorems to be extended to arbitrary relative Galois number fields.

Together with this outline, on which I shall enlarge later, I should like to illustrate the *principle* of the application of the noncommutative to the commutative: *By means of the theory of algebras, one tries to obtain invariant and simple formulations of known facts on quadratic forms or cyclic fields, i.e., such formulations that depend only on structure properties of*

Translated by Christina M. Mynhardt, Department of Mathematics, University of South Africa, Pretoria, South Africa, from *Verhandl. Intern. Math.-Kongress Zürich 1* (1932), 189-194.

the algebras. Once one has proven these invariant formulations—as is the case in the above-mentioned examples—these facts automatically carry over to arbitrary Galois fields.

2. Prior to a detailed description I should like to give a general survey of the different methods and further results. First it should be noted that the main difficulty lies in obtaining the formulation for general Galois fields, for which no starting point save the hypercomplex method is available at all; in the cited examples the underlying extension to the noncommutative has been obtained by the *simultaneous consideration of fields and groups* by means of the "crossed product" and its multiplication constants, the "factor systems" (cf. Section 3). In this way one obtains a simple normal algebra over the ground field and each such algebra is essentially obtainable thus. Such crossed products were first considerd by Dickson,[1] while the theory of factor systems was developed by Speiser, Schur, and R. Brauer[2] from a totally different starting point, namely, from the problem of absolutely irreducible representations. A sufficiently simple and far-reaching construction to handle commutative problems can be obtained only by combining both these theories.[3]

At the same time new and clear proofs for known facts have also been obtained: Here I should like to refer to a hypercomplex proof by H. Hasse[4] shortly to appear in *Math. Ann.*, of the reciprocity law for cyclic fields by means of an invariant formulation of his norm residue symbol, which is based on the theory of the crossed product. And further I refer to a hypercomplex foundation of local class field theory based on the same foundation recently given by C. Chevalley, in the course of which even more new algebraic theorems on factor systems were developed.[5] At the same time, however, I have to remark with reservation that the method of crossed products alone apparently does not yield the complete theory of Galois number fields. This follows from new and as yet unpublished results of Artin which are linked with the above cited proof of Hasse in the sense of the principle mentioned, but give only numerical equalities instead of complete isomorphism theorems.

One already has algebraic methods which yield a complete isomorphism—indeed, an operator isomorphism. It is a question of the continuation of beginnings by A. Speiser[6] and indeed of the interpretation of the Galois field as "Galois module," i.e., as a module over the ground field that admits the action of the Galois group as operators. And there exists an operator isomorphism between field and group ring (group algebra) in the sense that a unique correspondence between the elements occurs in such a way that linear forms correspond to the ground field and that the multiplication in the group

Hypercomplex Systems

ring corresponds to the action of the Galois group on the field. This theorem formulated by me[7] was proved by M. Deuring,[8] who based on it a proof of the Galois theory in which the operator isomorphism realizes the correspondence between group and field. Further structure theorems, likewise by Deuring, are analogous to the facts about Artin's L-series and yield a structural approach to the Artin conductors. These Artin L-series and conductors,[9] which are composed of general group characters, present apart from Speiser[6] the first connection between number theory and representation theory, a first advance over the abelian fields. They gave the whole development a strong impulse; the theory of Galois modules in particular acquired direction.

3. I should now like to discuss the problems, the norm principle and principal genus theorem mentioned in the beginning, in detail. First, the definition of the crossed product: Let K/k be a Galois field of degree n and $\mathfrak{G}$ its group. The crossed product denotes a simultaneous embedding of K and $\mathfrak{G}$ in an algebra A such that the automorphisms of K are inner. If the symbols $u_{S_1}, \ldots, u_{S_n}$ denote the n group elements, then one first writes A as a module of linear forms of rank n over k:

(1) $A = u_{S_1} K + \cdots + u_{S_n} K$ (i.e., A consists of all linear forms $u_{S_1} a_1 + \cdots + u_{S_n} a_n$ with a_i arbitrary in K).

By virtue of the requirement of the inner automorphisms (generated by the u_S, more generally, $u_S K^*$)[10], A becomes a ring, hence an algebra of rank n^2 over k. Namely, the requirement is expressed as

(2) $u_S^{-1} z u_S = z^S$ or [11] $z u_S = u_S z^S$ for each z in K.

(3) $u_S u_T = u_{ST} a_{S,T}$ with $a_{S,T}$ in K^*.

(4) $a_{ST,R} a_{S,T}^R = a_{S,TR} a_{T,R}$ (associative law from $[u_S u_T] u_R = u_S [u_T u_R]$).

A is called the crossed product of K with $\mathfrak{G}$, with factor system $a_{S,T}$. One proves that A is a simple normal algebra over k, hence a matrix ring D_r of degree r over the associated division algebra D, and that K is a maximal commutative subfield, hence a splitting field (i.e., extending the coefficient domain k with a field isomorphic to K yields a split algebra, a complete matrix ring over the center). Conversely, for any given division algebra D there are always matrix rings D_r which can be expressed as crossed products in the prescribed way.

Should one replace u_S with $v_S = u_S c_S$ with c_S in K^*, which generates the same automorphism, then "equivalent" factor systems result.

(5) $\bar{a}_{S,T} = a_{S,T} c_S^T c_T / c_{ST}$.

Equivalent factor systems are collected into a class (a); one likewise collects all algebras similar to A, i.e., all D_r with $r = 1, 2, \ldots$, into a class $\mathfrak{A}$. *The classes $\mathfrak{A}$ and (a) correspond one-to-one: the classes with fixed splitting field K form under tensor product an abelian group which is isomorphic to the term-by-term product of the classes of factor systems. The identity element is the class of split algebras or the system of all transformation elements* $c_S^T c_T / c_{ST}$. This is the algebra class group, first studied by R. Brauer.

4. By *specializing* to *cyclic splitting fields* I now want to come to the connection with the *norm concept* and in doing so, to the formulation of the generalized norm principle according to the principle explained at the start. If Z is cyclic and S a generating element of its group—the associated algebra is then called cyclic—then one can let the powers of S correspond to the powers of u; thus

(1′) $A = Z + uZ + \cdots + u^{n-1} Z$.

(2′) $Zu = uz^S$.

(3′) $u^n = \alpha$.

(4′) α lies in the ground field k^*.

(5′) $\bar{\alpha} = \alpha \cdot N(c)$, when $v = uc$.

Hence each factor system here consists of a single element α of the ground field [notation $A = (\alpha, Z)$]; the unit class of the factor systems is given by the *norms from* Z^*, and the group of algebra classes is isomorphic to the factor group $k^*/N(Z^*)$. Therefore, a cyclic algebra (α, Z) decomposes if and only if α is a norm of an element of Z. This relationship between norm and decomposition gives the formulation of the "norm principle," viz., the *theorem on the splitting of algebras: If an algebra splits at each place, then it splits completely.* As is customary in number theory the term "place" here is defined to mean that the ground field k is replaced by its $\mathfrak{p}$-adic extension k_p, where $\mathfrak{p}$ is a prime ideal of k (respectively, the finitely many infinite places corresponding to the embeddings of k and its conjugates in the field of real numbers).

This actually contains the norm theorem for cyclic fields. For according to the above, the theorem is, for cyclic algebras (α, Z), equivalent to the

Hypercomplex Systems

statement: if α is a $\mathfrak{p}$-adic norm at each (finite or infinite) place, then α is a norm of a number in Z, or, without passing to the $\mathfrak{p}$-adic: *If α is a norm residue for each prime ideal $\mathfrak{p}$ of k (and satisfies certain algebraic sign conditions), then α is a norm of a number in Z.* This latter formulation, however, is the norm principle proved in class field theory by using well-known analytical means. And the proof of the general theorem on splitting algebras can be obtained from this cyclic special case by pure algebraic-arithmetical considerations.[12] A first important corollary was found by Hasse: Each simple normal algebra over an algebraic number field is cyclic. The general formulation resulted from the search for a proof of this long-conjectured fact.

5. A second corollary of the theorem on splitting algebras—again with pure algebraic-arithmetical methods of reasoning—is the principal genus theorem[13] mentioned in the beginning. Its invariant formulation is based on the fact that the relations (2) to (5) defining the crossed product are purely multiplicative and therefore remain meaningful when K^* is replaced by an abelian group $\mathfrak{J}$ which merely satisfies the condition that its automorphism group contains a subgroup isomorphic to $\mathfrak{G}$. Relation (1) is then replaced by the "extension of $\mathfrak{G}$ with $\mathfrak{J}$" in the group-theoretical sense. If one takes the group of all ideals of K as $\mathfrak{J}$, then the factor systems become systems of ideals; a partition in $\mathfrak{J}$ induces a partition of the factor systems, and this is indeed in general a finer partition into ideal classes than the original. For the requirement that the multiplication of the u_S with (absolute) ideal classes be unique, means exactly that the transformation elements $c_S^T c_T / c_{ST}$—the unit class of the element factor systems—lie in the unit class of the ideal factor systems. Actually, however, as specialization to the known cases shows, a somewhat less fine partition suffices already. I define: *all those elements $a_{S,T}$ which generate split algebras at all (finite and infinite) ramified primes of K are considered to be in the unit class of the factor systems.* The resulting extension of $\mathfrak{G}$ is denoted by $\mathfrak{G}^*$; ($\mathfrak{c}$) denotes the absolute ideal class of $\mathfrak{c}$. Then one can state the *invariant formulation of the principal genus theorem:*

If an automorphism of $\mathfrak{G}^*$ results from the thusly defined partition by the substitution $v_S = u_S(\mathfrak{c}_S)$—all these $(\mathfrak{c}_S)$ form the principal genus—then the automorphism is inner and is generated by an ideal class ($\mathfrak{b}$).

The known special cases follow from an equivalent, somewhat more explicit formulation: *If the transformation elements $(c_S^T)(c_T)/(c_{ST})$ formed from the ideal classes ($\mathfrak{c}_S$) belong to the unit ideal class of the factor systems, then there exists an ideal class ($\mathfrak{b}$) such that the ($\mathfrak{c}_S$) are symbolic $(1-S)$th powers: $(\mathfrak{c}_S) = (\mathfrak{b})/(\mathfrak{b}^S)$ for all S in $\mathfrak{G}$.* For the assumption expresses exactly the automorphism property; the fact that this is an inner automorphism is expressed by $v_S = (\mathfrak{b})^{-1} u_S(\mathfrak{b}) = u_S(\mathfrak{b})^{1-S} = u_S(\mathfrak{c}_S)$.

The specialization to the cyclic case hence yields (under consideration of the normalization): If $N([\mathfrak{c}])$ lies in the unit ideal class of the factor systems, then $(\mathfrak{c})$ is a symbolic $(1-S)$th power: $(\mathfrak{c}) = (\mathfrak{b})^{1-S}$.

6. To proceed from here to the known situation for cyclic fields and quadratic forms, I remark first of all that although the theorem is expressed for complete ideal classes, its contents remain the same when one restricts oneself as usual to the ramified places of prime ideals.

In this way, however, the principal genus defined here goes over into the Gaussian because the ideal classes correspond to the quadratic forms, and the norms of the classes to the numbers representable by the forms. That the unit class generates the algebras which split at the ramified places of K can therefore be expressed by saying that these representable numbers are quadratic residues at the ramified places; the associated forms thus possess the total character of the principal form, hence form the Gaussian principal genus. It is already known that the $(1-S)$th symbolic power goes over into the square.

For cyclic fields the formulation becomes as follows: The principal genus consists of all ideal classes whose norms are norm residues at the (finite and infinite) ramified places. That, however, is of course equivalent to "norm residue with respect to the conductor" and from this the usual theorem results by specialization.

Furthermore, in the general case of arbitrary Galois fields, one can also introduce a conductor composed of the ramified places only, such that in normalizing the factor systems for each of these places, the ray contains only elements of the unit class.

Here the question of the relationship to the Artin conductors mentioned in Section 2, which are composed of the same prime ideals, arises, and thereby the question of the relationship of the second hypercomplex method to the theory of Galois modules. How far these two methods will reach, the future will still have to show.

NOTES

1. Cf. his book *Algebren und ihre Zahlentheorie,* Zürich, 1927, §34.
2. Cf. for example, R. Brauer, Untersuchungen über die arithmetischen Eigenshaften von Gruppen linearer Substitutionen, *Math. Ztschr. 28* (1928), and the literature given there in Remark 2.

3. I initially developed this construction in a lecture in winter 1929-1930, reproduced in Chapter 2 of H. Hasse, Theory of cyclic algebras, *Transac. 134* (1932). A report of M. Deuring on hypercomplex numbers and number-theoretical applications which will appear in *Ergebnisse der Mathematik* treats the whole field covered in the lecture.
4. H. Hasse, Die Struktur der R. Brauerschen Algebrenklassengruppe über einem algebraischen Zahlkörper (insbesondere Normenrest-symbol und Reziprozitätsgesetz), *Math. Ann. 107* (1932-1933).
5. C. Chevalley, Sur la théorie du symbole de restes normiques, *J. f. Math. 169*.
6. A. Speiser, Gruppendeterminante und Körperdiskriminante, *Math. Ann. 77* (1916).
7. E. Noether, Normalbasis bei körpern ohne höhere Verzweigung (Satz 3), *J. f. Math. 167* (1932) (the proof contained an error).
8. M. Deuring, Galoissche Theorie und Darstellungstheorie, *Math. Ann. 107* (1932).
9. E. Artin, Über eine neue Art von L-Reihen, *Math. Sem. Hamburg 3* (1924). Zur Theorie der L-Reihen mit allgemeinen Gruppencharacteren, ibid. *8* (1931). Die gruppentheoretische Struktur der Diskriminanten algebraischer Zahlkörper, *J. f. Math. 164* (1931).
10. K^* results from K by omission of the zero; this notation is generally used.
11. z^S means as usual the element obtained from z by the substitution S.
12. R. Brauer, H. Hasse, E. Noether, Beweis eines Hauptsatzes in der Theorie der Algebren, *J. f. Math. 167* (1932).
13. The proof will appear in *Math. Ann.*

Publications
of Emmy Noether

1. Über die Bildung des Formensystems der ternären biquadratischen Form, *Sitz. Ber. d. Physikal.-mediz. Sozietät in Erlangen 39* (1907), 176-179.
2. Über die Bildung des Formensystems der ternären biquadratischen Form, *Jour. f. d. reine u. angew. Math. 134* (1908), 23-90 and two tables.
3. Zur Invariantentheorie der Formen von n Variabeln, *Jahresber. Deutsch. Math.-Verein 19* (1910), 101-104.
4. Zur Invariantentheorie der Formen von n Variabeln, *Journ. f. d. reine u. angew. Math. 139* (1911), 118-154.
5. Rationale Funktionkörper, *Jahresber. Deutsch. Math.-Verein 22* (1913), 316-319.
6. Körper und Systeme rationaler Funktionen, *Math. Annalen 76* (1915), 161-191.
7. Der Endlichkeitssatz der Invarianten endlicher Gruppen, *Math. Annalen 77* (1915), 89-92.
8. Über ganze rationale Darstellung der Invarianten eines Systems von beliebig vielen Grundformen, *Math. Annalen 77* (1915), 93-102.
9. Die allgemeinsten Bereiche aus ganzen transzendenten Zahlen, *Math. Annalen 77* (1916), 103-128. Corrig., *Math. Annalen 81*, 30.
10. Die Funktionalgleichungen der isomorphen Abbildung, *Math. Annalen 77* (1916), 536-545.
11. Gleichungen mit vorgeschriebener Gruppe, *Math. Annalen 78* (1918), 221-229. Corrig., *Math. Annalen 81*, 30.

12. Invarianten beliebiger Differentialausdrücke, *Nachr. d. König. Gesellsch. d. Wiss. zu Göttingen, Math-phys. Klasse* (1918), 38-44.
13. Invariante Variationsprobleme, *Nachr. d. König. Gesellsch. d. Wiss. zu Göttingen, Math-phys. Klasse* (1918), 235-257.
14. Die arithmetische Theorie der algebraischen Funktionen einer Veränderlichen in ihrer Beziehung zu den übrigen Theorien und zu der Zahlkörpertheorie, *Jahresber. Deutsch. Math.-Verein. 28, Ab. 1* (1919), 182-203.
15. Die Endlichkeit des Systems der ganzzahligen Invarianten binärer Formen, *Nachr. d. König Gesellsch. d. Wiss. zu. Göttingen, Math-phys. Klasse* (1919), 138-156.
16. Zur Reihenentwicklung in der Formentheorie, *Math. Annalen 81* (1920), 25-30.
17. With W. Schmeidler: Moduln in nichtkommutativen Bereichen, insbesondere aus Differential- und Differenzenausdrücken, *Math. Zeitschr. 8* (1920), 1-35.
18. Über eine Arbeit des im Kriege gefallenen K. Hentzelt zur Eliminationstheorie, *Jahresber. Deutsch. Math.-Verein. 30, Ab. 2* (1921), 101.
19. Idealtheorie in Ringbereichen, *Math. Annalen 83* (1921), 24-66.
20. Ein algebraisches Kriterium für absolute Irreduzibilität, *Math. Annalen 85* (1922), 26-33.
21. Formale Variationsrechnung und Differentialinvarianten, *Encyklopädie d. math. Wiss. III, 3, E* (1922), 68-71. (in: R. Weitzenböck, Differentialinvarianten).
22. Based on the dissertation of Kurt Hentzelt, who died before this paper was presented. Zur Theorie der Polynomideale und Resultanten, *Math. Annalen 88* (1923), 53-79.
23. Algebraische und Differentialinvarianten, *Jahresber. Deutsch. Math.-Verein. 32* (1923), 177-184.
24. Eliminationstheorie und allgemeine Idealtheorie, *Math. Annalen 90* (1923), 229-261.
25. Eliminationstheorie und Idealtheorie, *Jahresber. Deutsch. Math.-Verein. 33* (1924), 116-120.
26. Abstrakter Aufbau der Idealtheorie im algebraischen Zahlkörper, *Jahresber. Deutsch. Math.-Verein. 33* (1924), 102.
27. Hilbertsche Anzahlen in der Idealtheorie, *Jahresber. Deutsch. Math.-Verein. 34, Ab. 2* (1925), 101.
28. Gruppencharaktere und Idealtheorie, *Jahresber. Deutsch. Math.-Verein, 34, Ab. 2* (1925), 144.
29. Ableitung der Elementarteiler theorie aus der Gruppentheorie, *Jahresber. Deutsch. Math.-Verein 34, Ab. 2* (1926), 104.
30. Der Endlichkeitssatz der Invarianten endlicher linearer Gruppen der Charakteristik p, *Nachr. d. Gesellsch. d. Wiss. zu Göttingen, Math.-phys. Klasse* (1926), 28-35.

31. Abstrakter Aufbau der Idealtheorie in algebraischen Zahl- und Funktionskörpern, *Math. Ann. 96* (1926), 26-61.
32. Der Diskriminantensatz für die Ordnungen eines algebraischen Zahl- oder Funktionenkörpers, *Jour. f. d. reine u. angew. Math. 157* (1927), 82-104.
33. With R. Brauer: Über minimale Zerfällungskörper irreduzibler Darstellungen, *Sitzungsber. d. Preuss. Akad. d. Wiss.* (1927), 221-228.
34. Hyperkomplexe Grössen und Darstellungstheorie, in arithmetischer Auffassung, *Atti Congresso Bologna 2* (1928), 71-73.
35. Hyperkomplexe Grössen und Darstellungstheorie, *Math. Zeitschr. 30* (1929), 641-692.
36. Über Maximalbereiche von ganzzahligen Funktionen, *Rec. Soc. Math. Moscou 36* (1929) 65-72.
37. Idealdifferentiation und Differente, *Jahresber. Deutsch. Math.-Verein. 39, Ab. 2* (1929), 17.
38. Normalbasis bei Körpern ohne höhere Verzweigung, *Jour. f. d. reine u. angew. Math. 167* (1932), 147-152.
39. With R. Brauer and H. Hasse: Beweis eines Hauptsatzes in der Theorie der Algebren, *Journ. f. d. reine u. angew. Math. 167* (1932), 399-404.
40. Hyperkomplexe Systeme in ihren Beziehungen zur kommutativen Algebra und Zahlentheorie, *Verhandl. Internat. Math. Kongress Zürich 1* (1932), 189-194.
41. Nichtkommutative Algebren, *Math. Zeitschr. 37* (1933), 514-541.
42. Der Hauptgeschlechtssatz für relativ-galoissche Zahlkörper, *Math. Annalen 108* (1933), 411-419.
43. Zerfallende verschränkte Produkte und ihre Maximalordnungen, Exposés mathématiques publiés à la mémoire de J. Herbrand IV, *Actualités scient. et. industr. 148* (1934).
44. Idealdifferentiation und Differente, *Jour. f. d. reine u. angew. Math. 188* (1950), 1-21.

Index

Alexandroff, P. S., 17, 22, 24-25, 71
Algebraic number theory, 157-163
Artin, E., 68, 71, 85, 150
Ascending chain condition, 133

Bernays, P., 70, 72, 77
Brauer, R., 152-153, 170
Brauer-Hasse-Noether theorem, 153-154

Calculus of variations, 125-130
Class field theory, 160-162
Commutative ring theory, 131-143
Courant, R., 18, 24-25, 67, 74

Dedekind, R., 21, 27-28, 148
Dedekind domain, 138
Deuring, M., 40, 90

Einstein, A., 13, 34-35, 45-46

Fischer, E., 12

Fitting, H., 40-41
Frobenius, G., 148-149

Galois theory, 115-124
Gordan, P., 5, 7-8
Group representations
  history of, 147-150
  in physics, 147

Hasse, H., 44
Herbrand, J., 43
Herglotz, G., 41, 68, 77
Hilbert, D., 10-11, 13-14, 20, 22, 66-67, 77, 145, 158-159, 161
Hopf, H., 22, 25, 110

Integral dependence, 136

Klein, F., 4, 13, 20, 145
Krull, W., 84, 141

Landau, E., 68-69
Levitski, J., 40

Lie group theory, 130

Mac Lane, S., 44
McKee, R. S., 89
Molien, T., 149-150
Mori, S., 141-142

Noether, A., 5
Noether, E., 77
  in America, 30-39
  anecdotes about, 24-25, 70-71, 83
  at 1932 ICM, 27
  curriculum vitae of, 15
  death of, 37-38, 91
  dissertation of, 8-10
  influence on algebraic topology, 71-72, 110
  reference letters for, 34-36
  relationship with O. Taussky, 81-82, 87-89
  Soviet ties, 108-111
  students of, 12, 28, 39-43, 89
Noether, F., 5
Noether, M., 4, 6-7, 20
Noetherian ring, 134-135
Noncommutative ring theory, 145

Primary decomposition, 134-135, 142
Principal genus theorem, 162-163, 169-171
Representation theory, 145-156

Schilling, O., 41
Scott, C., 31
Shoda, K., 42, 91
Steenrod's problem in topology, 123

Taussky, O., 33, 66, 71
Thomas, M., 32
Tsen, C., 41

van der Waerden, B., 19, 21, 23-24, 85, 106, 141, 145
Weber, W., 28-29, 40, 69
Weyl, H., 23, 69-70, 77
Wheeler, A., 32, 86, 89-90
Witt, E., 41, 70
Women
  Noether's views on, 91
  status of, 9-10, 14